Sustainable Civil Infrastructures

Editor-in-Chief

Hany Farouk Shehata, SSIGE, Soil-Interaction Group in Egypt SSIGE, Cairo, Egypt

Advisory Editors

Khalid M. ElZahaby, Housing and Building National Research Center, Giza, Egypt

Dar Hao Chen, Austin, TX, USA

Sustainable Civil Infrastructures (SUCI) is a series of peer-reviewed books and proceedings based on the best studies on emerging research from all fields related to sustainable infrastructures and aiming at improving our well-being and day-to-day lives. The infrastructures we are building today will shape our lives tomorrow. The complex and diverse nature of the impacts due to weather extremes on transportation and civil infrastructures can be seen in our roadways, bridges, and buildings. Extreme summer temperatures, droughts, flash floods, and rising numbers of freeze-thaw cycles pose challenges for civil infrastructure and can endanger public safety. We constantly hear how civil infrastructures need constant attention, preservation, and upgrading. Such improvements and developments would obviously benefit from our desired book series that provide sustainable engineering materials and designs. The economic impact is huge and much research has been conducted worldwide. The future holds many opportunities, not only for researchers in a given country, but also for the worldwide field engineers who apply and implement these technologies. We believe that no approach can succeed if it does not unite the efforts of various engineering disciplines from all over the world under one umbrella to offer a beacon of modern solutions to the global infrastructure. Experts from the various engineering disciplines around the globe will participate in this series, including: Geotechnical, Geological, Geoscience, Petroleum, Structural, Transportation, Bridge, Infrastructure, Energy, Architectural, Chemical and Materials, and other related Engineering disciplines.

SUCI series is now indexed in SCOPUS
and EI Compendex.

More information about this series at http://www.springer.com/series/15140

Hany Shehata · Sherif El-Badawy
Editors

Sustainable Issues in Infrastructure Engineering

The official 2020 publication
of the Soil-Structure Interaction Group
in Egypt (SSIGE)

 Springer

Editors
Hany Shehata
CEO of SSIGE
Cairo, Egypt

Sherif El-Badawy
Mansoura University
Mansoura, Egypt

ISSN 2366-3405   ISSN 2366-3413   (electronic)
Sustainable Civil Infrastructures
ISBN 978-3-030-62585-6   ISBN 978-3-030-62586-3   (eBook)
https://doi.org/10.1007/978-3-030-62586-3

This Springer imprint is published by the registered company Springer Nature Switzerland AG
The registered company address is: Gewerbestrasse 11, 6330 Cham, Switzerland

Contents

SSIGE Official Publications 2020: Part 2

FE Modeling of RC Beams Reinforced in Flexure with BFRP Bars Exposed to Harsh Conditions

Hakem Alkhraisha, Haya Mhanna, and Farid Abed[✉]

Civil Engineering, American University of Sharjah, Sharjah, UAE
fabed@aus.edu

Abstract. This paper aim to investigate the effect harsh conditions have on the flexural behavior of concrete beams reinforced longitudinally with Basalt Fiber Reinforced Polymer (BFRP) bars. Finite element (FE) software, ABAQUS, is used to develop a nonlinear model capable of simulating the behavior of exposed FRP reinforced beams in flexure. Data extracted from the numerical simulations were compared with experimental data to validate the FE model. Through a parametric study, this paper aims to study the effect of reducing the modulus of elasticity on BFRP reinforced beams. The values of modulus of elasticity are reduced from the by 10%, 20%, and 30% and the impact is observed. The paper also presents comparative analysis among different BFRP tensile strength values. The analysis of the effect of varying tensile strength values on the behavior of BFRP RC beams will be conducted on an under-reinforced beam. The BFRP tensile strength value is reduced by 10%, 20%, and 30% and the impact is observed. The results show that the reduction of the modulus of elasticity of the BFRP bar decreased the flexural capacity of the BFRP RC beam proportionally. The proportional decrease was not affected by the number of reinforcement bars used in the beams with similar axial stiffness. In addition, a reduction the tensile strength of the BFRP bars caused a disproportional decrease in the flexural capacity of the beams. Beams with lower tensile strength values failed at lower deflections.

1 Introduction

Fiber Reinforced Polymer (FRP) reinforcement was created to combat as an alternative to the conventional steel reinforcements. FRP products are non-metallic material which makes then non-corrosive (ACI 440 2015). Furthermore, FRP composites possess larger tensile strength than steel (ACI 440 2015; Bedard 1992). FRP reinforcements have different chemical composition and exhibit different failure modes than steel reinforcements do. Therefore, the conventional design philosophies of reinforced concrete had to be altered to account for the different mechanical behavior of FRP reinforcement. FRP reinforcement does not have unified material properties. Their properties depend on fiber type, fiber volume, fiber orientation, resin type, and quality control during the development process (ACI 440 2015). Despite the discrepancies in properties, all types of FRP bars are made of high strength fibers such as basalt, carbon, glass or aramid, combined together by a polymer resin (ACI 440 2015; Bedard 1992).

FRP bars are anisotropic materials that show linear elastic behavior until rupture (440 2015; Bedard 1992). Furthermore, FRP bars have lower modulus of elasticity than

© The Author(s), under exclusive license to Springer Nature Switzerland AG 2021
H. Shehata and S. El-Badawy (Eds.): *Sustainable Issues in Infrastructure Engineering*, SUCI, pp. 3–13, 2021.
https://doi.org/10.1007/978-3-030-62586-3_1

steel (440 2015; El Messalami et al. 2019 Bedard 1992; Hollaway 2003). The ultimate tensile strength of FRP bars is indirectly related to their size; in other words an increase in diameter decreases the tensile strength (Hollaway 2003).

BFRP reinforcement, being a more recent development, has fewer studies about its flexural and shear behaviors (Abed and Alhafiz 2019; Abed et al. 2019; El Refai and Abed 2016; Elgabbas et al. 2017; El Refai et al. 2015; Fam and Tomlinson 2015; Abed et al. 2012). Elgabbas et al. studied the behavior of over-reinforced beams under static loads. The results showed that the increase in reinforcement ratio caused a non-linear increase in the flexural capacity of the beams (Elgabbas et al. 2017). Tomlinson et al. studied the performance of beams reinforced with BFRP in flexure and shear. Nine beams with reinforcement ratio ranging from 0.28 to 1.60 were casted. The results showed that the ultimate and service loads were proportionally impacted by the reinforcement ratio. The mode of failure of the nine beams depended on the reinforcement ratio and shear reinforcement (Fam and Tomlinson 2015).

Some researchers investigated the effect of exposure on the bond behavior between BFRP bars and concrete and on the flexural behavior of BFRP RC beams. The studies varied from examining the mechanical behavior of exposed FRP bars to investigating the effects of exposure on BFRP RC beams. Furthermore, studies had investigated the effects of Alkaline solutions (Al Rifai et al. 2020; Sim et al. 2005; Wang et al. 2008; Wu et al. 2015), temperature (El Refai et al. 2014; Calvet et al. 2015; D'Antino et al. 2018; Masmoudi et al. 2011; Robert et al. 2009), and saline solutions (Altalmas et al. 2015; Yan and Lin 2017; Al-Tamimi et al. 2014).

Wu et al. studied tensile properties of BFRP bars exposed to alkaline solution, salt solution, acid solution, and deionized water (Wu et al. 2015). The study utilized scanning electronic microscopy (SEM) to monitor the degradation mechanism of BFRP bars in an alkaline environment. The results indicated that acid, salt, and deionized water had a lower impact on the durability of BFRP bars than the alkaline solution did (Wu et al. 2015). Sim et al. examined the durability of basalt, glass, and carbon fiber exposed to alkali solution combined with high temperatures (Sim et al. 2005). The results showed that after seven days exposure, basalt and glass FRPs lost 50% of their strength and volume. Carbon FRP, however, only lost 13% of its strength during the same exposure time. The resin and fiber type played a big role in degradation resistance and since GFRP and BFRP had almost the same chemical composition they degraded at similar rate. The CFRP, however, did not face any relatively significant degradation. Wang et al. had also studied the durability of BFRP exposed to alkali solution (Wang et al. 2008). After three months of exposure the strength decrease by 40% but the modulus of elasticity remained unaffected. The results also indicated BFRP had lower degradation than the basalt fiber with no resin protection (Wang et al. 2008).

Calvet et al. investigated the effect of the environmental conditions on the bond behavior of different CFRP bars. The study indicated that at high temperatures the textured part of the CFRP bar separated from the bar (Calvet et al. 2015). CFRP had shown a better bond behavior than the reference steel bars did. The bond strength was 10% higher at the CFRP reinforcement system than at the steel (Calvet et al. 2015). Another study that investigated bond behavior was conducted by Altalmas et al. in 2015. The study analyzed the bond degradation of BFRP bars exposed to accelerated aging conditions. The bars were exposed to aggressive environment such as acid, saline, and alkaline. The results indicated that despite exposure, BFRP bars showed higher bond strengths to concrete than the ribbed GFRP did. The surface texture controlled the slippage resistance, as sand coated bars had better adhesion to concrete than ribbed bars regardless of the fiber type. GFRP bars exposed to acid solution suffered 25% loss in strength while the ones exposed to alkali and sea water lost 17% of its original strength (Altalmas et al. 2015).

Robert et al. investigated the behavior of GFRP reinforcing bars subjected to extreme temperatures. The study found that temperatures from 40° to 50 °C there was no significant effect on the tensile strength and modulus of elasticity (Robert 2010). Masmoudi et al. studied the effect of temperature ranging from 20 °C to 80 °C on the bond performance of GFRP bars in concrete. Eighty pull out specimens were tested to determine the bond behavior of the GFRP bars. The exposure period reached up to eight months and the highest temperature reached 80 °C. the study found that the exposure cause a 14% reduction in bond strength of the tested GFRP specimens (Masmoudi et al. 2011).

Research about the impact of exposure on BFRP bars using finite element analysis is limited. The main objective of this research is, therefore, to provide a FE model that can simulate the behavior of BFRP reinforced beams. The model would, then, provide much needed data about the flexural response of BFRP RC beams impacted by exposure. Over-reinforced beams would be used to examine the effect of varying BFRP bars modulus of elasticity on the flexural behavior of BFRP RC beams. Additionally, under-reinforced beams would be used to examine the effect of varying BFRP bars tensile strength on the flexural behavior of BFRP RC beams.

2 Brief Overview of the Experimental Program

The experimental program consisted of four RC beams reinforced with different configurations of BFRP bars that were exposed to UAE climate to investigate the effect of exposure to harsh environments on the capacity of the beams. The tested beams are 2200 mm long, 230 mm deep, and 180 mm wide with a clear span of 1900 mm as shown in Fig. 1. All beams were reinforced by $\phi 8$ stirrups spaced at 100 mm c/c to force flexural failure. The main flexural BFRP reinforcement are 2 ϕ 8, 2 ϕ 12, 2 ϕ 16 and 3 ϕ 12 for each beam, respectively. In addition, the tested specimens were reinforced with 2 ϕ 10 steel bars in the compression zone. Table 1 shows the tested beams designations and description. It should be noted that the beam specimens were loaded using two-point loading test at a displacement control mode rate of 1 mm/min.

The average concrete compressive strength (f'c) was found by averaging the values of crushing three cylinders. The average f'c of the beam specimens in this study is 36 MPa. In addition, uniaxial coupon tensile tests were conducted to measure the elastic modulus and tensile strength of BFRP and steel reinforcement. The average elastic modulus and tensile strength of BFRP bars was 59 GPa and 1200 MPa, respectively. In addition, the average Young's modulus and tensile strength of the steel bars was 200 GPa and 460 MPa, respectively.

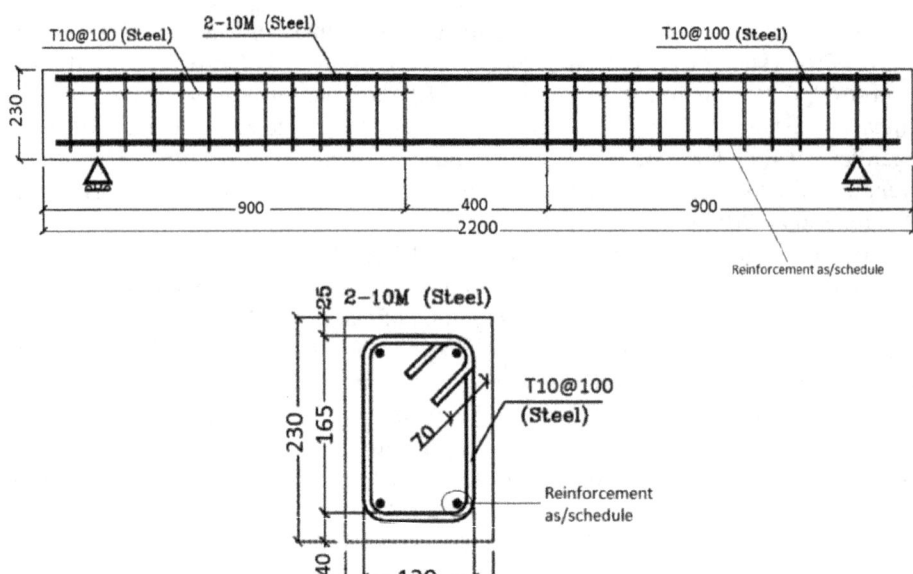

Fig. 1. Reinforcement details

Table 1. Test matrix

Beam designation	Description
2T8B	2 # 8 BFRP bars
2T12B	2 # 12 BFRP bars
3T12B	3 # 12 BFRP bars
2T16B	2 # 16 BFRP bars

2.1 Finite Element Modeling (FEM)

All beams investigated in this study were simulated using commercial finite element software (ABAQUS). The geometric and materials nonlinearities were captured by incorporating (*NLGEOM) parameter in the general static step. Full 3D beams were modeled and the load-displacement responses of four beams (2T8B, 2T12B, 3T12B, and 2T16B) were compared to the experimental results to validate the FE models. Following that, a parametric study was conducted to investigate the effect of reduction of modulus of elasticity of BFRP for over-reinforced beams (3T12B and 2T16B). In addition, the effects of reduction in tensile strength of BFRP for under-reinforced beams (2T6B) were investigated. Table 2 presents the modulus of elasticity and tensile strength of BFRP used to modeled beams in this study.

Table 2. FEM matrix

Beam designation	Modulus of elasticity (GPa)	Tensile strength (MPa)	Failure mode
Group 1 (3T12B; reduced modulus of elasticity)			
3T12B	53.1	1200	Concrete crushing
3T12B	47.2	1200	Concrete crushing
3T12B	41.3	1200	Concrete crushing
Group 2 (2T16B; reduced modulus of elasticity)			
2T16B	53.1	1200	Concrete crushing
2T16B	47.2	1200	Concrete crushing
2T16B	41.3	1200	Concrete v
Group 3 (FRP type)			
2T6B	59	1200	FRP rupture
2T6B	59	1080	FRP rupture
2T6B	59	960	FRP rupture
2T6B	59	840	FRP rupture

2.2 Material Properties

For the purpose of this study, three materials were defined in ABAQUS: concrete, steel and BFRP. The complex behavior of concrete was characterized by using concrete damage plasticity model provided by the software to simulate the nonlinearity of concrete material. Details of the parameters incorporated in concrete damage plasticity model are available in Abed et al. (2020). Compressive and tensile stress-strain curves that were input in the model were obtained based on the compressive strength of concrete as shown in Fig. 2.

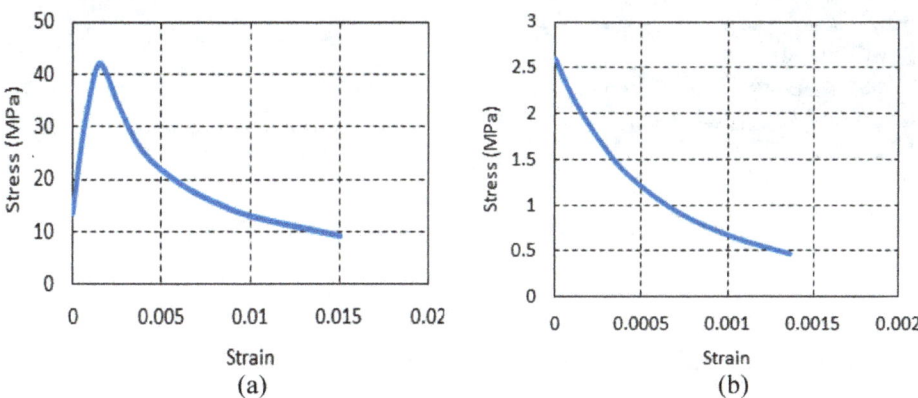

Fig. 2. Concrete material properties: (a) Compressive stress-strain curve; (b) Tensile stress-strain curve

Steel material, used in top reinforcement and stirrups, was defined to have an elastic modulus of 200 GPa and Poisson's ratio of 0.3. The plastic property, i.e. yield stress, was defined to be 460 MPa.

The behavior of BFRP bars is linear elastic, until it reaches its tensile capacity, after which it ruptures without yielding. All BFRP bars were modeled to have an elastic modulus of 59 GPa and Poisson's ratio 0.2. In over-reinforced beams the tensile capacity was defined to be 1200 MPa at zero plastic strain because failure is dominated by concrete crushing. On the other hand, the plastic property of under-reinforced beams was modeled to have a tensile strength of 1200 MPa at zero plastic strain.

2.3 Beam Geometry and Mesh Sensitivity

Steel and FRP bars were modeled as 2-node linear 3-D truss elements (T3D2) that are embedded in the concrete region. The concrete part was modeled as an 8-node linear 3D solid element with reduced integration (C3D8R). Mesh sensitivity was carried out to select the optimum mesh size that is not considered course and simulates the experimental results with minimum computational time. In this analysis, a mesh size of 30 mm was considered ideal for all beams. Figure 3 shows the reinforcement cage for a beam with 2 BFRP bars and mesh configuration for the beams modeled in this study. Four rigid plates were introduced to the model to apply onto the boundary conditions and reduce stress concentration on the beam in those areas. The interaction between the plates and the beam was modeled as surface-to-surface contact, with the beam being the slave surface and the plates as master surface. The boundary conditions at the bottom plates were a pin and a roller. In addition, a vertical displacement was applied at the top plates to mimic the applied load on the beam.

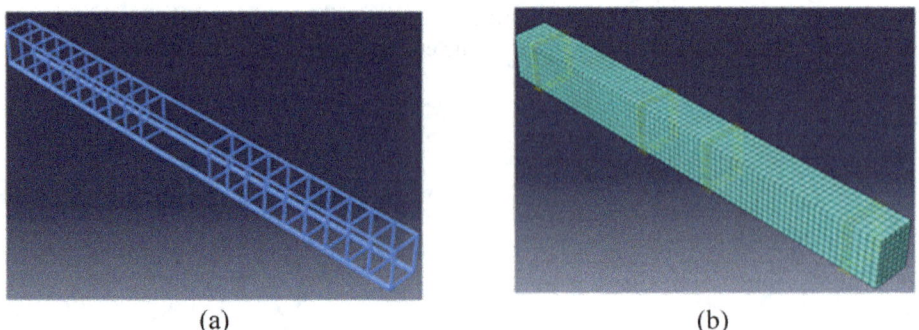

(a)   (b)

Fig. 3. FEM model: (a) Reinforcement cage; (b) Mesh configuration

2.4 Model Verification

The validation of FE models was conducted by comparing load-midspan deflection responses of the experimental results and FE generated curves. Figure 4 shows the tested and simulated load-displacement curves for beams 2T8B, 2T12B, 2T16B, and 3T12B, respectively. It is can be seen from Fig. 4 that all simulated results are in good agreement with the experimental results at all stages of loading until failure. Mainly, the load at first crack, initial stiffness, and ultimate capacity of the beams maximum load and displacement of the FE models aligns with the experimental curves.

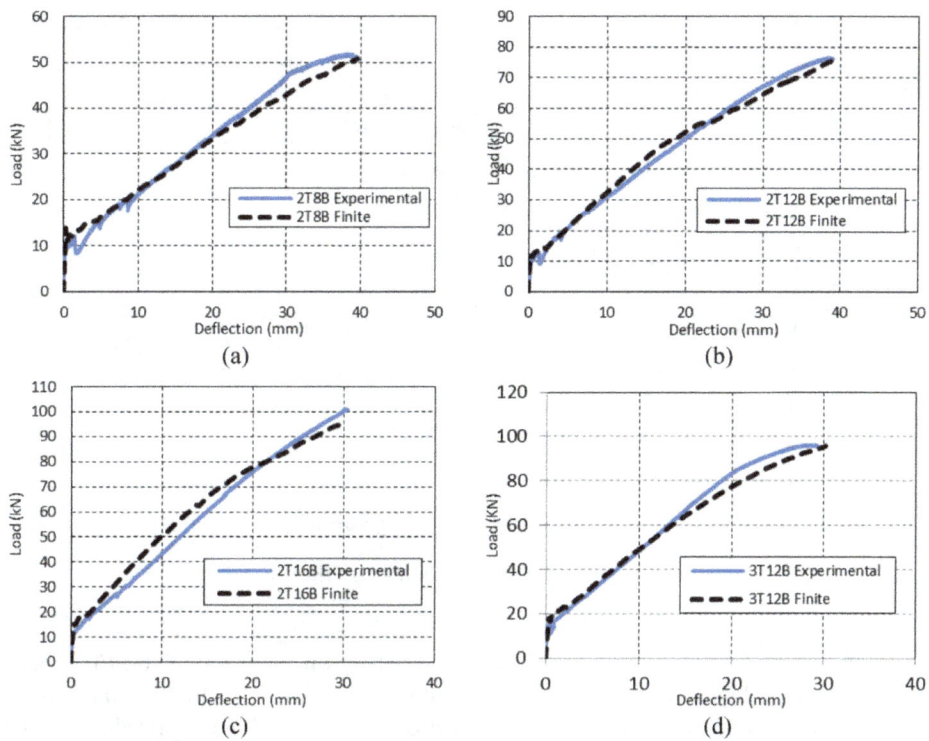

Fig. 4. Model verification for beams: (a) 2T8B; (b) 2T12B; (c) 2T16B; (d) 3T12B.

3 Discussion of Results

The verified models were utilized to carry out two parametric studies. Variations in the modulus of elasticity and tensile strength of BFRP were implemented and the flexural behavior of the varying models was observed.

Figure 5 presents the load versus midspan deflection behavior of the FE models examining the effect of reduction of the modulus of elasticity of BFRP on the flexural behavior of 3T12B beam. The modulus of elasticity values used were 1200 MPa, 1080 MPa, 960 MPa, and 840 or 0% 10%, 20%, and 30% reduction, respectively. The reduction of the modulus of elasticity caused a proportional reduction in the flexural capacity of the beam. The reduction becomes visible, only, after the first crack load which can be attributed to the similar concrete properties of the beam. Concrete properties control the behavior of the beam up until the first crack. After which, the properties of the BFRP controls the beams behavior.

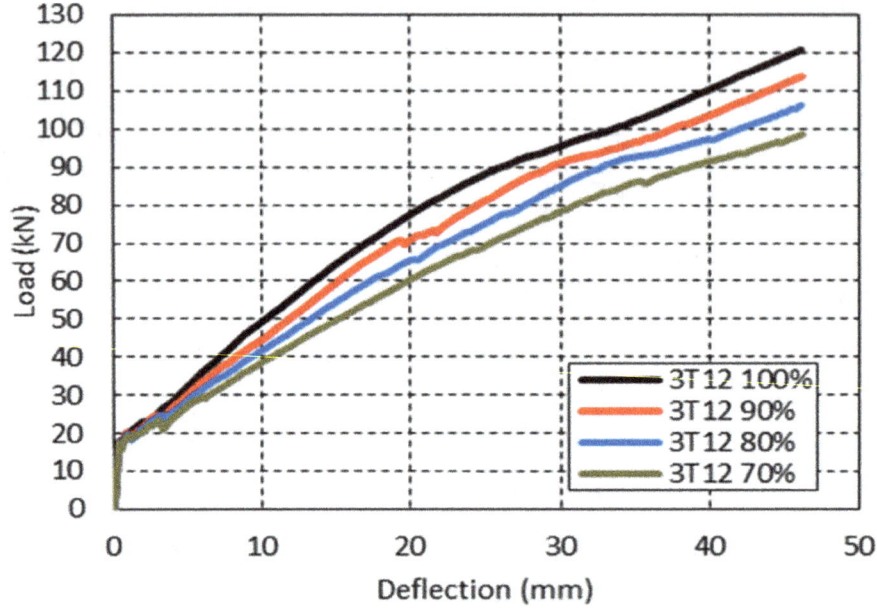

Fig. 5. Load-deflection curves for GROUP 1

Figure 6 presents the load versus midspan deflection behavior of the FE models examining the effect of reduction of the modulus of elasticity of BFRP on the flexural behavior of 2T16B beam. The modulus of elasticity values used the same values used in the 3T12B model. The reduction, also, showed a proportional reduction of the flexural capacity of the beam due to the decrease in the modulus of elasticity of the BFRP. Figure 5 and Fig. 6 show that beams with similar axial stiffness react to exposure in a similar fashion despite the detailing of the beam. The reduction in flexural capacity is not affected by the number of BFRP bars present in the tension zone.

Figure 7 presents the load versus midspan deflection behavior of the FE models examining the effect of reduction of the tensile strength of BFRP on the flexural behavior of 2T6B (under-reinforced) beam. The reduction in tensile strength caused a disproportional decrease in the flexural capacity of the beam. A reduction of 10% (as compared to the beam with 0% reduction), for example, impacted the capacity of the beam more than a reduction of 80% (as compared to the beam with 10% reduction.

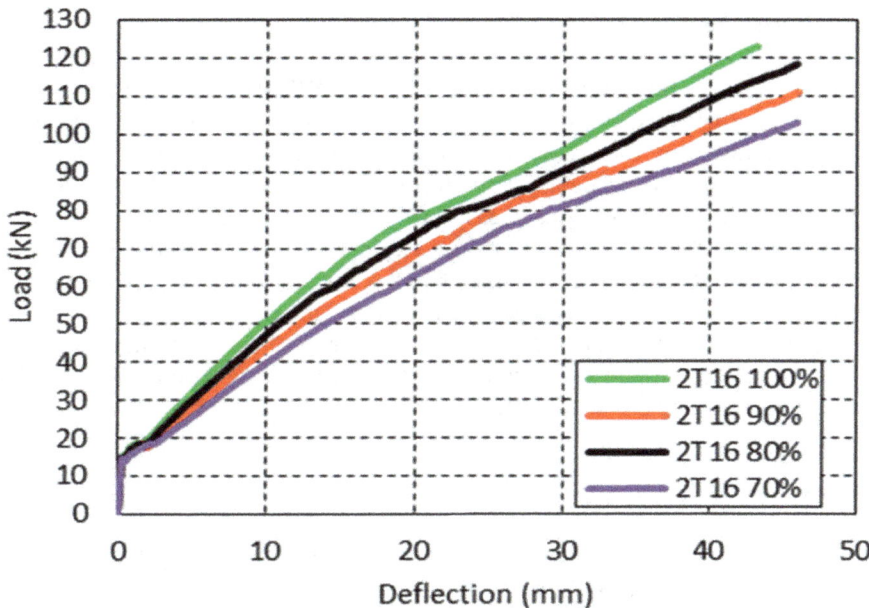

Fig. 6. Load-deflection curves for Group 2

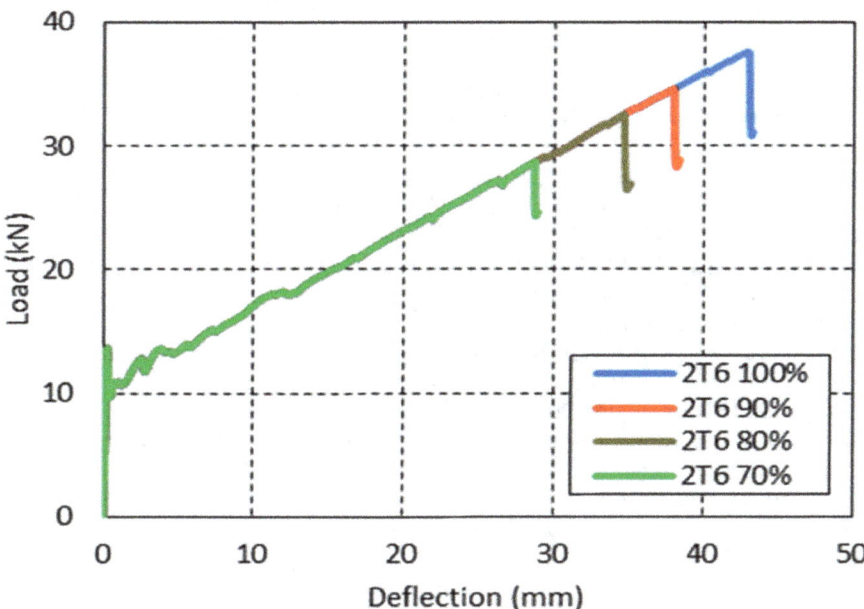

Fig. 7. Load-deflection curves for Group 3

4   Conclusions

After verifying the validity of the FE model using four BFRP RC beams, three nonlinear FE models were developed to study the response of RC beams reinforced in flexure with BFRP bars affected by exposure. Parametric analysis included variation in modulus of elasticity and tensile strength was performed. From the FE results, the following conclusions could be drawn. A decrease in the modulus of elasticity of BFRP bars (caused by exposure to harsh conditions) caused a proportional reduction in the flexural capacity of BFRP RC beams. The reduction in flexural capacity of the BFRP RC beams due to the reduction of the modulus of elasticity of the beams was not impacted by the detailing of the beam (i.e. number of reinforcement bars) of beams with similar axial stiffness. Reduction in the tensile strength of BFRP bars caused a disproportional decrease in the flexural capacity of BFRP RC beams.

References

ACI 440: Guide for the Design and Construction of Structural Concrete Reinforced with Fiber-Reinforced Polymer Bars. Farmington Hills, MI (2015)

Abed, F., Oucif, C., Awera, Y., Mhanna, H., Alkhraisha, H.: FE modeling of concrete beams and columns reinforced with FRP composites. Defense Technol. (2020, in press). https://doi.org/10.1016/j.dt.2020.02.015

Abed, F., El Refai, A., Abdalla, S.: Experimental and finite element investigation of the shear performance of BFRP-RC short beams. Structures **20**, 689–701 (2019). https://doi.org/10.1016/j.istruc.2019.06.019

Abed, F., Alhafiz, A.R.: Effect of basalt fibers on the flexural behavior of concrete beams reinforced with BFRP bars. Compos. Struct. **215**, 23–34 (2019). https://doi.org/10.1016/j.compstruct.2019.02.050

Abed, F., El-Chabib, H., AlHamaydeh, M.: Shear characteristics of GFRP-reinforced concrete deep beams without web reinforcement. J. Reinf. Plast. Compos. **31**, 1063–1073 (2012). https://doi.org/10.1177/0731684412450350

Al Rifai, M., El-Hassan, H., El-Maaddawy, T., Abed, F.: Durability of basalt FRP reinforcing bars in alkaline solution and moist concrete. Constr. Build. Mater. **243**, 118258 (2020). https://doi.org/10.1016/j.conbuildmat.2020.118258

Altalmas, A., El Refai, A., Abed, F.: Bond degradation of basalt fiber-reinforced polymer (BFRP) bars exposed to accelerated aging conditions. Constr. Build. Mater. **81**, 162–171 (2015). https://doi.org/10.1016/j.conbuildmat.2015.02.036

Al-Tamimi, A., Abed, F., Al-Rahmani, A.: Effects of harsh environmental exposures on the bond capacity between concrete and GFRP reinforcing bars. Adv. Concr. Constr. **2**(1), 1 (2014). https://doi.org/10.12989/acc.2014.2.1.001

Bedard, C.: Composite reinforcing bars: assessing their use in construction. Concr. Int. **14**, 55–59 (1992)

Calvet, V., Valcuende, M., Benlloch, J., Cánoves, J.: Influence of moderate temperatures on the bond between carbon fibre reinforced polymer bars (CFRP) and concrete. Constr. Build. Mater. **94** 589–604 (2015)

D'Antino, T., Pisani, M.A., Poggi, C.: Effect of the environment on the performance of GFRP reinforcing bars. Compos. B **141**, 123–136 (2018). https://doi.org/10.1016/j.compositesb.2017.12.037

El Messalami, N., El Refai, A., Abed, F.: Fiber-reinforced polymers bars for compression rein-forcement: a promising alternative. Constr. Build. Mater. **209**, 725–737 (2019). https://doi.org/10.1016/j.conbuildmat.2019.03.105

El Refai, A., Abed, F.: Concrete contribution to shear strength of beams reinforced with basalt fiber-reinforced bars. J. Compos. Constr. **20** (2016).https://doi.org/10.1061/(asce)cc.1943-5614.0000648

El Refai, A., Abed, F., Altalmas, A.: Bond durability of basalt fiber–reinforced polymer bars embedded in concrete under direct pullout conditions. J. Compos. Constr. **19**(5), 04014078 (2014). https://doi.org/10.1061/(ASCE)CC.1943-5614.0000544

El Refai, A., Abed, F., Al-Rahmani, A.: Structural performance and serviceability of concrete beams reinforced with hybrid (GFRP and steel) bars. Constr. Build. Mater. **96**, 518–529 (2015). https://doi.org/10.1016/j.conbuildmat.2015.08.063

Elgabbas, F., Benmokrane, B., Ahmed, E.: Flexural behavior of concrete beams reinforced with ribbed basalt-FRP bars under static loads. J. Compos. Constr. **21** (2017). https://doi.org/10.1061/(asce)cc.1943-5614.0000752

Fam, A., Tomlinson, D.: Performance of concrete beams reinforced with basalt FRP for flexure and shear. J. Compos. Constr. **19** (2015). https://doi.org/10.1061/(asce)cc.1943-5614.0000491

Hollaway, L.C.: The evolution of and the way forward for advanced polymer composites in the civil infrastructure. Constr. Build. Mater. **17**, 365–378 (2003). https://doi.org/10.1016/S0950-0618(03)00038-2

Masmoudi, R., Masmoudi, A., Ouezdou, M.B., Daoud, A.: Long-term bond performance of GFRP bars in concrete under temperature ranging from 20 °C to 80 °C. Constr. Build. Mater. **25**, 486–493 (2011)

Robert, M.: Behavior of GFRP reinforcing bars subjected to extreme temperatures. J. Compos. Constr. **14**, 353–360 (2010). https://doi.org/10.1061/(ASCE)CC.1943-5614.0000092

Robert, M., Cousin, P., Benmokrane, B., Canadian Society for Civil Engineering Annual Conference St. Johns NC: Behaviour of GFRP reinforcing bars subjected to extreme temperatures. In: Proceedings, Annual Conference - Canadian Society for Civil Engineering, vol. 3, pp. 1587–1596

Sim, J., Park, C., Moon, D.Y.: Characteristics of basalt fiber as a strengthening material for concrete structures. Compos. B **36**, 504–512 (2005). https://doi.org/10.1016/j.compositesb.2005.02.002

SIMULAI: Abaqus/CAE User's Manual, Rising Sun Mills, RI (2019)

Wang, M., Zhang, Z., Li, Y., Li, M., Sun, Z.: Chemical durability and mechanical properties of alkali-proof basalt fiber and its reinforced epoxy composites. J. Reinf. Plast. Compos. **27**, 393–407 (2008)

Wu, G., Dong, Z.-Q., Wang, X., Zhu, Y., Wu, Z.-S.: Prediction of Long-Term Performance and Durability of BFRP Bars under the combined effect of sustained load and corrosive solutions. J. Compos. Constr. **19**, 04014058 (2015) . https://doi.org/10.1061/(ASCE)CC.1943-5614.0000517

Yan, F., Lin, Z.: Bond durability assessment and long-term degradation prediction for GFRP bars to fiber-reinforced concrete under saline solutions. Compos. Struct. **161**, 393–406 (2017). https://doi.org/10.1016/j.compstruct.2016.11.055

Manufacturing and Mechanical Testing of Casuarina Glauca Blockboards

Nour Abdellatif[1], Mohamed Darwish[2](✉), Khaled Nassar[1], Passant Youssef[1], Abdallah Dardir[1], Abdulrahman Ahmed[1], Mahmoud Eltamimy[1], Mohamed Mamdouh[1], and Rami Abdelazim[1]

[1] Department of Construction Engineering, American University in Cairo, Cairo, Egypt
[2] Department of Construction and Building Engineering, Alahram Canadian University, American University in Cairo, Cairo, Egypt
mdarwish@aucegypt.edu

Abstract. Wood, or timber, is light, cheap, and easy to transport and work with. On the contrary, reinforced concrete is more expensive, heavy & difficult to transport and slower to build. Previous research has proven that relatively cheap Casuarina Glauca wood could have sufficiently high strength that makes it a strong candidate for structural usage. In the construction field in Egypt, most of the used wood is blockboard composed from imported wood. The importing of wood represents a significant segment and Egypt invests a lot of its money in this segment, thus a research such as this one would significantly help Egypt to save money in the import business. This research aims to produce and test Egyptian blockboard made from Casuarina Glauca wood farmed in Egypt. The blockboards were produced and tested for their mechanical properties and compared to their imported counterparts. Moreover, these blockboards proved to be of a sufficiently high strength. Based on the results, this engineered wood product could be a structurally sound alternative for structural usage.

1 Introduction

Based on the research done, the Egyptian construction market consumption of wood is draining the Egyptian market from a lot of investment and is considered costly since most of the wood products in Egypt is imported. Egypt started importing wood since 1957 and right now the total imports of wood costs approximately $1.57 billion USD. Which is about 2% of the total exports in Egypt. Using such mass of money in investing in the Egyptian market and enhancing the wood industry in Egypt which currently have high quality resources as seen in country report. Egypt is trying to advance in the forestation sector, thus giving manufacturers the ability to use such products from the Egyptian Agriculture. Egypt is planting 3 tropical trees in Upper Egypt named Khaya, Teak and Neem. Also some water treatment plants in Egypt grows Casuarina trees near wastewater since such trees can endure and survive using wastewater. This shows how the forestation in Egypt is in a phase of expansion. Nowadays, The Seed Bank that is developed in the past years in Egypt carries tests and experiments new techniques in breeding trees in Egypt and also stores seeds and plants coming abroad from other countries. This ensures

H. Shehata and S. El-Badawy (Eds.): *Sustainable Issues in Infrastructure Engineering*, SUCI, pp. 14–22, 2021.
https://doi.org/10.1007/978-3-030-62586-3_2

the continuity of the improvement in the Egyptian wood forestation. Nowadays, Egypt is in a shortage of producing wood (FAS 2015).

Meanwhile a previous study by Hussein et al. (2019a) have covered the mechanical properties of different types of Casuarina. Within that study, the Glauca specie of Casuarina was proven to be even stronger than several types of oaks. The study was further extended to study the different properties of that wood within different percentages of moisture content Hussein et al. (2019b).

The high strength of that type of wood encouraged Mahmoud et al. (2019) to further study its mechanical properties, its grading categorization and the usage of coatings to protect it. Furthermore, Mahmoud et al. (2019b) have studied its usage to design a 12-m span truss and have experimentally proven that the truss members made of Casuarina Glauca could carry even more than the design loads.

Several researchers like Zanuttini and Cremonini (2002) and Haseli et al. (2018) have studied the manufacturing of blockboards but unfortunately none of them have manufactured blockboards of sufficient strength using cost-effective raw materials like Casuarina Glauca. Furthermore, other researchers like Youssef et al. (2019) have used Egyptian wood waste to produce an engineered wood product but again it was not as strong as the parent wood itself.

The current research focuses on using the locally available and relatively strong Casuarina Glauca wood in manufacturing blockboards. Consequently, the Egyptian blockboard produced in this research could fill a lot of the gaps as it uses the locally available, cost-effective and strong Casuarina Glauca wood to produce a structurally sound and cost-effective blockboard. This was done through manufacturing several thicknesses of the blockboard, testing them till failure and comparing their results with the market-available imported counterparts. Furthermore, a cost comparison between the locally manufactured and imported blockboards was performed. The Egyptian blockboard proved to be more sound in terms of strength and cost-effectiveness and would be a promising candidate to be used in building structures as to be seen in this paper.

2 Manufacturing and Testing Procedures

2.1 Production of Casuarina Glauca BlockBoard

The process of blockboard production could be summarized into eight main steps which are:

1. Obtaining the Casuarina Glauca wood from the supplier and obtaining the veneer wood.
2. Preparation of the wood, profiling the wood so that it has a smooth surface.
3. Sawing the wood into blocks or battens with a dimension of (35 * 18 * 400 mm) with a clean saw and this is crucial for two reasons the first is that when the battens are arranged next to each other no voids would be present between them that would weaken the blockboard and the second is for a better cohesion between the battens.
4. Arranging the wood in a brick-like pattern so that we don't have any weak planes inside of the board by not having two joints next to each other as shown in Fig. 1.

Fig. 1. The arrangement of wood within a blockboard panel.

Fig. 2. The sandwiching of wood in between the two veneer layers.

5. Formulating the glue, the glue that was used was a glue consisting of a risen and a hardener. The risen was Phenol formaldehyde and the hardener was ammonia.
6. Equally spreading the glue on the veneer and checking not to miss a space.
7. Taking the blocks or battens with the arrangement done in step 4 and placing them on the veneer and placing the other layer of the glued veneer, with this step we have a sandwich like structure with a veneer on top and bottom as shown in Fig. 2.
8. Applying heat and pressure on the block board so that the glue wood hardens and gain a higher strength and no voids exists between the battens and the veneer. The heat and pressure was applied for 12 min using the hot-press device and then the board would be removed.

2.2   Testing of Casuarina Glauca Blockboard

In construction, wood in most likely used as a compression member or a member to resist bending, it is not likely to see wood used as a member to carry a tension force or torsion, thus when testing the blockboard compression and the 4-point bending tests were performed. Compression testing was performed twice; parallel to the wood grain and perpendicular to the wood grain. Two different orientations for the 4-point bending were tested; one orientation was the load being applied to the thickness dimension thus the resisting section was the thickness of the blockboard and the other orientation was the specimen width was the resisting section. The sizing of the specimens for the bending samples the specimens size was 18 mm * 92.5 mm * 400 mm which allowed a clear span of 310 mm and for the compression samples the specimens size was 18 mm * 92.5 mm * 133 mm. In the compression tests the specimens of the blockboard had a dimensioning of 18 mm * 100 mm * 130 mm where the 130 mm is the height and the 100 mm is the width and the 18 mm is the thickness. The load was applied on the 100 mm * 18 mm surface when the compression test was parallel to the grain as shown in Fig. 3a. Meanwhile, the load was applied on the 130 mm * 18 mm surface when the compression test was perpendicular to the grain as shown in Fig. 3b. The specimen was done in these dimensions to avoid buckling thus testing for compressive strength of the wood. The testing of these specimens was done on a MTS machine and according to the ASTM and the referred ASTM tests were D143-14, D4761-19 with some deviations from the ASTM. In the compression and bending tests, the specimen tested was placed in the MTS and a metal disc was placed on the sample as to insure that the load would be well distributed along the specimen. In the perpendicular compression test the load was applied at a rate of 2 mm/min while in the parallel compression test the rate was 4 mm/min the reason for increasing the rate was due to knowing that the wood is anisotropic material and can carry more loads in the direction of parallel to the fiber. In the 4-point bending test the rate was kept constant 2 mm/min regardless of the orientation.

a. Testing for compression parallel to the grains

b. Testing for compression perpendicular to the grains

Fig. 3. The experimental set-up for the compression tests.

3   Results and Discussion

The results of the bending tests shown in Fig. 4 and Fig. 5 for the two different bending orientations show that the blockboard samples made out of Glauca wood were stronger in bending than their counterparts made from imported wood. Furthermore, the stress-strain diagrams of the Glauca blockboards were more consistent with small deviations between them however, their imported wood counterparts had significant variations between them which could be attributed to the variety of sources that these woods were imported from while the Glauca samples were all from the same farm. Moreover, in the Casuarina sample the catastrophic failure was in the veneer layer or the outer layer and the core or the battens were intact and this would promote the Casuarina Blockboard in two ways the first is that these cores are reusable and the second reason is it would serve as an early warning system such as that of the reinforced concrete and its ductile mode of failure.

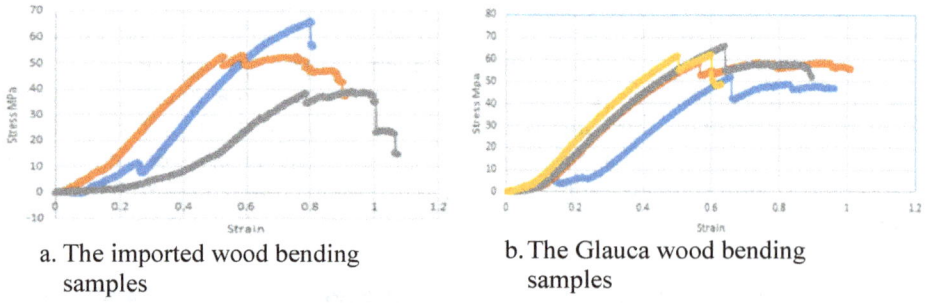

a. The imported wood bending   b. The Glauca wood bending
   samples   samples

Fig. 4. The stress-strain curves for the bending test with 92 mm depth.

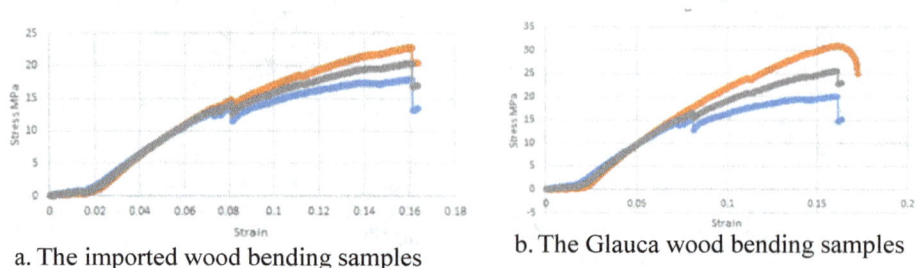

a. The imported wood bending samples   b. The Glauca wood bending samples

Fig. 5. The stress-strain curves for the bending test with 18 mm depth.

The results of the compression parallel to the grain tests shown in Fig. 6 show that the blockboard samples made out of Glauca wood were stronger in bending than their counterparts made from imported wood. The results of the compression perpendicular to the grain tests shown in Fig. 7 show that the blockboard samples made out of Glauca wood were stronger in bending than their counterparts made from imported wood. Also and similar to the behavior in bending, the stress-strain diagrams of the Glauca blockboards were more consistent with small deviations between them however, their imported wood counterparts had significant variations between them which could be attributed to the variety of sources that these woods were imported from while the Glauca samples were all from the same farm.

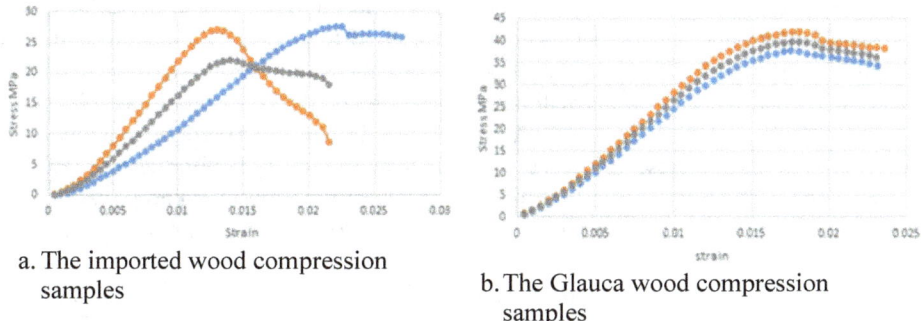

a. The imported wood compression samples

b. The Glauca wood compression samples

Fig. 6. The stress-strain curve for the compression parallel to the grain tests

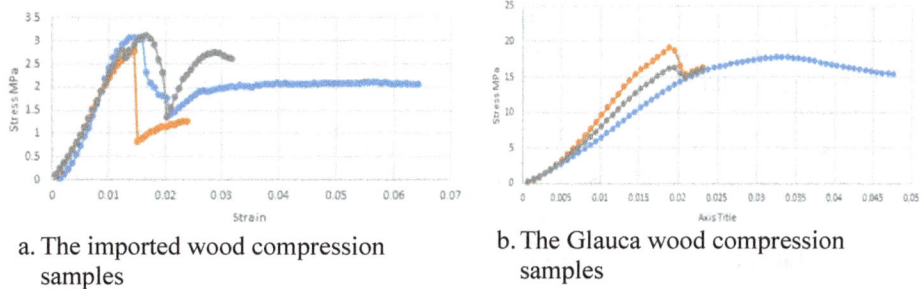

a. The imported wood compression samples

b. The Glauca wood compression samples

Fig. 7. The stress-strain curves for the compression perpendicular to grain tests.

4   Cost Analysis and Comparison

Moreover, a comparison of a cost breakdown structure of the Egyptian blockboard manufactured by the authors and the imported blockboard where both blockboards are with a thickness of 18 mm was done. In this study the comparison between the boards was a comparison between cores as the only difference between the two boards, though we had the option of obtaining Egyptian veneer however that was avoided as not to have too many variables. Thus in calculating the costs the only difference between the board manufactured by us and the imported one is the core cost. Thus in the table above you can see that the equipment cost and labor cost doesn't change this is because the imported block board uses the same equipment and labor cost. Moreover, the only aspect under the scope is the core and thus when we came to calculate the cost, we needed to know how much volume was in a single board and this calculation was obtained by volume of wood per one board is equal to the thickness of the core inside the board multiplied by the length of the board multiplied by the width of the board which resulted in a volume of 0.053 cubic meters per board which is a constant between the two board. The difference was in the unit volume price of the wood cores, the imported wood also known as Swedish redwood or pine was priced at 5000 Egyptian pounds per cubic meter while the Casuarina Egyptian wood was priced at 3500 Egyptian pounds per cubic meter. In calculating the labor cost, two artisans are needed to produce the blockboard. One worker will arrange the wood in a wall brick like pattern in order not to have any weak planes in the board and the other will be preparing the wood for installation in the board as mentioned in part A of the report. A time of 20 min was measured for installation of the wood in the board theses 20 min are including an efficiency factor thus a production of 24 boards per day assuming that a working day is 8 h. Now we know that we can produce 24 boards per days we will assume that the worker who is preparing the wood works with the same rate and thus 24 boards multiplied by 0.053 this will result in 1.3 cubic meters which is the amount of wood needed to be sawn per day for production. Now we can calculate the costs, we know that a skilled labor daily wage is 300 Egyptian pounds and we know the cost of sawing wood is 600 Egyptian pounds per cubic meter. Therefore, the first workers cost per board is his daily wage divided by the number of boards he produces which will result in a 12.5 EGP/board. The second worker cost is calculated by multiplying the cost of sawing wood by the volume per board which results in a 32.5 EGP/board thus in total labor cost is 45 EGP/board. Lastly the equipment cost is calculated through calculating the electricity bills the depreciation and maintenance needed cost.

A summary of the cost comparison is shown in Table 1. These results prove the cost-effectiveness of the proposed system and its possible usage on an industrial stage.

Table 1. Cost of Casuarina Glauca blockboard vs imported blockboard

Direct cost of a single blockboard (2.44 m*1.22 m*0.018 m)		
	Egyptian blockboard 18 mm	Imported blockboard 18 mm
Materials	Veneer = 200LE Blockboard core = 187 LE Binder = 3 LE	Veneer = 200LE Blockboard core = 267 LE Binder = 3 LE
Labor	12.5 + 32.5 = 45 LE	45 LE/board
Equipment	200 LE	200 LE
Total direct cost	615 LE	695 LE

5   Conclusions and Recommendations

Based on the experimental work performed together with the cost analysis, the following conclusions could be drawn:

- The blockboards made out of Casuarina Glauca were stronger in bending than their imported counterparts.
- The blockboards made out of Casuarina Glauca were stronger in compression than their imported counterparts.
- The blockboards made out of Casuarina Glauca were more consistent in their mechanical properties than their imported counterparts.
- The blockboard made of Casuarina Glauca are more cost-effective when compared to their imported counterparts.

In lieu of the performed study, it is recommended to:

- Study the structural performance of the blockboard made of Casuarina Glauca on a large scale.
- Study the cost-effectiveness of the blockboard made of Casuarina Glauca on an industrial level.
- Study the different manufacturing options related to the veneer used in blockboard manufacturing.

Acknowledgments. The authors would like to acknowledge the lab personnel in the department of construction engineering at the American university in Cairo for their continuous help during this study.

References

ASTM Standard D143-14: Standard Test Methods for Small Clear Specimens of Timber. ASTM International, West Conshohocken, PA. www.astm.org

ASTM Standard D4761-19: Standard Test Methods for Mechanical Properties of Lumber and Wood-Based Structural Materials. ASTM International, West Conshohocken, PA. www.astm.org

FAS: Egypt's Wood Sector Report (2015). https://www.fas.usda.gov/data/egypt-egypts-wood-sector-report

Haseli, M., Layeghi, M., Hosseinabadi, H.Z.: Characterization of block board and batten board sandwich panels from date palm waste trunks. Measurement **124**, 329–337 (2018)

Hussein, M., Nassar, K., Darwish, M.: Mechanical properties of Egyptian casuarina wood. ASCE J. Mater. Civ. Eng. **31**(12) (2019a). https://doi.org/10.1061/(ASCE)MT.1943-5533.0002955

Hussein, M., Darwish, M., Nassar, K.: The effect of moisture content on some mechanical properties of Casuarina wood. In: Proceedings of The Seventh International Conference on Structural Engineering, Mechanics and Computation (SEMC 2019), Cape Town, South Africa (2019b)

Mahmoud, A., Fayez, M., ElSayed, M., Said, J., Nadeem, S., Yacoub, O., Nassar, K., Yazeed, E., Darwish, M., AbouZeid, M., AbouAli, R.: Material testing of Casuarina Glauca wood for usage within a truss. In: CSCE Annual Conference, Laval, Canada (2019)

Mahmoud, A., Said, J., Bader, S., Sayed, M., Fayez, M., Yacoub, O., Youssef, P., Darwish, M., Nassar, K., Sayed-Ahmed, E.Y., AbouZeid, M.N.: Design and testing of pre-engineered prefabricated casuarina wooden truss. In: Proceedings of the Seventh International Conference on Structural Engineering, Mechanics and Computation (SEMC 2019), Cape Town, South Africa (2019)

Youssef, P., Zahran, K., Nassar, K., Darwish, M., Elhaggar, S.: Manufacturing of wood-plastic composite boards and their mechanical and structural characteristics. ASCE J. Mater. Civ. Eng. **31**(10) (2019). https://doi.org/10.1061/(ASCE)MT.1943-5533.0002881

Zanuttini, R., Cremonini, C.: Optimization of the test method for determining the bonding quality of core plywood (Blockboard). Mater. Struct. **35**(2), 126–132 (2002). https://doi.org/10.1007/bf02482112

Comparing Shear Strength Prediction Models of Ultra-High-Performance Concrete Girders

Antony Kodsy and George Morcous$^{(\boxtimes)}$

University of Nebraska-Lincoln, 1110 S. 67th Street, Omaha, NE 68182-0816, USA
akodsy@unomaha.edu, gmorcous2@unl.edu

Abstract. The use of Ultra-High-Performance Concrete (UHPC) in bridge construction has been growing rapidly in the last 15 years due to its excellent mechanical and durability properties. One of the areas of interest to bridge engineers is the elimination of transverse reinforcement in precast/prestressed concrete girders as it simplifies girder fabrication and result in smaller and lighter girder sections. UHPC has a relatively high post-cracking tensile strength due to the presence of steel fibers, which enhance its shear strength and eliminate the need for transverse reinforcement. In this paper, three prediction models for the shear strength of UHPC are discussed: RILEM TC 162-TDF 2003, *fib* Model Code 2010, and French Standard NF P 18-710 2016. Data obtained from several UHPC shear experiments in the literature was collected to evaluate the prediction models of the shear strength of UHPC girders. Comparing predicted versus measured shear strength of UHPC girders indicated that the French Standard NF P 18-710 2016 model provides the closest prediction, while the fib Model Code 2010 model provides the least scattered prediction of UHPC shear strength.

Keywords: UHPC · Shear · French Standard 2016 · RILEM · *fib* Model code

1 Introduction

Recently, Ultra-High-Performance Concrete (UHPC) gained significant popularity in bridge construction which titled it to be called a "Game Changer" in the field, Binard 2017. Having exceptional mechanical properties compared to conventional concrete, UHPC enables the production of relatively shallow, slender, and light girders without transverse reinforcement. The presence of steel fibers in UHPC girders enhances the post-cracking tensile strength that often controls the shear strength in beams (i.e. diagonal tension). Several experimental investigations have been conducted to quantify the shear strength of UHPC beams when transverse reinforcement is eliminated. This paper evaluates the shear strength prediction models of UHPC beams by comparing the measured versus predicted strengths of shear test data. The shear test data was collected from fifteen experimental shear test programs done on UHPC beams over the past two decades.

© The Author(s), under exclusive license to Springer Nature Switzerland AG 2021
H. Shehata and S. El-Badawy (Eds.): *Sustainable Issues in Infrastructure Engineering*, SUCI, pp. 23–39, 2021.
https://doi.org/10.1007/978-3-030-62586-3_3

2 Prediction Models

Few code prediction models currently exist to estimate the shear strength of UHPC girders, and relatively high safety margins are used in these models due to the lack of test data Baby et al. 2013a, b, and Graybeal 2006.

2.1 RILEM TC 162-TDF 2003

The Eurocode 2 part 1, 1991 shear strength prediction model of conventional concrete was used as the general framework to develop this prediction model Vandewalle 2000a, b. However, Eurocode 2 considers only the pre-peak behavior of concrete in tension, while this model considers the effect of steel fibers on the post-peak behavior of fiber-reinforced concrete. Therefore, the stress-strain relationship of fiber-reinforced concrete is needed for this model. The ultimate shear load carrying capacity (V_{Rd3}) is taken to be the sum of the contributions of concrete (V_{cd}), and of the stirrups and/or inclined bars (V_{wd}), and the steel fibers (V_{fd}). Steel fibers contribution is calculated according to the following equations:

$$V_{fd} = 0.7 \, k_f k \, b_w \, d \, \tau_{fd} \quad [N] \tag{1}$$

Where (k_f) is a factor to account for the contribution of flanges in T-shaped sections (taken as 1.0 for other shapes) calculated as follows:

$$k_f = 1 + n\left(\frac{h_f}{b_w}\right)\left(\frac{h_f}{d}\right) \tag{2}$$

$$n = \frac{b_f - b_w}{h_f} \leq 3; \text{ and } n \leq \frac{3b_w}{h_f} \tag{3}$$

Where (b_f) is the width of flanges [mm]; (b_w) is the beam minimum web width [mm] over the effective depth (t); (d) is the effective depth [mm]; (h_f) is the height of flanges [mm]. And, (k) is the size effect factor taken as $1 + \sqrt{\frac{200}{d}} \leq 2$; ($\tau_{fd}$) is the design value of the increase in shear strength due to steel fibers taken as ($0.12f_{R,4}$) [MPa]; ($f_{R,4}$) is the residual flexural tensile strength corresponding to crack mouth opening displacement (CMOD) of 3.5 mm [MPa]. Residual flexural tensile strengths ($f_{R,i}$) are determined experimentally by a three-point bending test on a 150 × 150 × 550 mm (5.9 × 5.9 × 21.6 in.) notched prism. The stress-strain relation is obtained from the load-deflection or load-CMOD of the notched prism. The load-CMOD curve is defined by four points ($i = 1$ through 4) corresponding to CMOD of 0.5, 1.5, 2.5, and 3.5 mm, respectively. ($f_{R,i}$) is calculated as follows:

$$f_{R,i} = \frac{3F_{R,i}xL}{2bxh_{sp}^2} \quad [MPa] \tag{4}$$

Where ($F_{R,i}$) is the load recorded at crack mouth opening displacement ($CMOD_i$) [N]; (L) is the span of the prism [mm]; (b) is the width of the prism cross-section [mm]; (h_{sp})

is the distance between the tip of the notch to the top of the prism cross-section [mm]. Concrete contribution is calculated as follows:

$$V_{cd} = \left[0.12 \, k \, (100 \, \rho_1 \, f_{ck})^{\frac{1}{3}} + 0.15 \, \sigma_{cp}\right] b_w \, d \quad [N] \tag{5}$$

$$\rho_1 = \frac{A_l}{b_w d} \tag{6}$$

$$\sigma_{cp} = \frac{N_{sd}}{A_c} \quad [MPa] \tag{7}$$

Where (ρ_1) is the longitudinal reinforcement ratio (recommended not to exceed 2%); (A_l) is the area of tension reinforcement extending not less than (d + anchorage length) beyond the section considered [mm^2]; (f_{ck}) is the characteristic cylinder compressive strength [MPa]; (σ_{cp}) is the level of axial loading or prestressing in the section [MPa]; (N_{Sd}) is the longitudinal force in the section due to loading or prestressing (compression: positive) [N]; (A_C) is the cross-sectional area of the beam [mm^2].

2.2 *fib* Model Code 2010

The prediction model is developed for steel fiber reinforced concrete with conventional strength and is not validated yet for UHPC. The approach used has a close resemblance to RILEM TC 162-TDF 2003 prediction model except that the steel fibers contribution to the resistance is coupled with the concrete contribution in one term. After cracking, the concrete contribution is weakly coupled with the transverse reinforcement contribution and is strongly coupled with the steel fibers contribution to the shear resistance, Foster 2018. The coupled contributions of concrete and fibers to the shear resistance of fiber reinforced concrete elements without transverse reinforcement is given by:

$$V_{Rd,F} = \left\{\frac{0.18}{\gamma_c} \cdot k \cdot \left[100 \cdot \rho_1 \left(1 + 7.5\frac{f_{Ftuk}}{f_{ctk}}\right) \cdot f_{ck}\right]^{1/3} + 0.15 \cdot \sigma_{cp}\right\} \cdot b_w \cdot d \quad [N] \tag{8}$$

Where terms like (k), (ρ_1), (σ_{cp}), (b_w), and (d) are defined similarly to RILEM TC 162-TDF 2003. (γ_c) is the partial safety factor for concrete without fibers typically taken as 1.5. However, (γ_c) was not considered when calculating the ultimate shear resistance of UHPC beams in the evaluation of the prediction model in this paper. (f_{Ftuk}) is the characteristic value of the ultimate residual tensile strength determined as follows:

$$f_{Ftuk} = \frac{f_{R,3}}{3}[MPa] \tag{9}$$

$$f_{Ftuk} = f_{Fts} - \frac{w_u}{CMOD_3}\left(f_{Fts} - 0.5f_{R,3} + 0.2f_{R,1}\right) \geq 0 \; [MPa] \tag{10}$$

$$f_{Fts} = 0.45 f_{R,1} \; [MPa] \tag{11}$$

Similar to RILEM TC 162-TDF 2003, ($f_{R,1}$) and ($f_{R,3}$) are the residual flexural tensile strengths corresponding to CMOD of 0.5 mm and 3.5 mm respectively [MPa]; (f_{Fts}) is the characteristic service residual tensile strength (post-cracking strength at serviceability crack opening) [MPa]; (w_u) is the maximum crack opening accepted in structural design (typically taken as 1.5 mm). The residual flexural tensile strengths are determined experimentally according to EN 14651 2007 by a three-point bending test on a 150 × 150 × 550 mm (5.9 × 5.9 × 21.6 in.) notched prism. (f_{Ftuk}) can be determined based on the rigid-plastic model Eq. (9), or the linear model based on Eqs. (10) and (11). (f_{ctk}) is the characteristic value of the tensile strength for the concrete without fibers [MPa] determined as follows:

$$f_{ctk} = 2.12 \ln(1 + 0.1(f_{ck} + 8\,MPa))[MPa] \quad for\,f_{ck} > 50\,MPa) \tag{12}$$

Where (f_{ck}) is the characteristic value of cylinder compressive strength [MPa].

2.3   French Standard, NF P 18-710 2016

Similar shear prediction model to the AFGC 2013, the shear resistance is equal to the smaller of the resistance of concrete compressive struts $(V_{Rd,max})$ and the tensile resistance of the ties (V_{Rd}). The general philosophy is similar to the AASHTO LRFD simplified version of the Modified Compression Field Theory procedure dividing (V_{Rd}) into concrete contribution term $(V_{Rd,c})$ and shear reinforcement contribution term $(V_{Rd,s})$, with the addition of fibers contribution term $(V_{Rd,f})$. The concrete contribution term for prestressed beam sections is calculated as follows:

$$V_{Rd,c} = \frac{0.24}{\gamma_{cf}\,\gamma_E} k f_{ck}^{1/2} b_w\, z \text{ [N]} \tag{13}$$

$$k = 1 + \frac{3\,6_{cp}}{f_{ck}}, \text{ for } 6_{cp} \geq 0 \tag{14}$$

The terms (k), (f_{ck}), (b_w), and (6_{cp}) are similar to what was described in the previous models. The terms (γ_{cf}) and (γ_E) are partial safety factors (typically taken as 1.5 after being multiplied). (γ_{cf}) is a partial factor for UHPC under tension typically taken as 1.3, while (γ_E) is a partial factor accounting for the uncertainty about extrapolating the model developed for high performance concrete with $(f_{ck} \leq 90\,MPa\,(13.1\,ksi))$ to UHPC, (z) is the internal moment lever arm (typically taken as 90% of the section depth). A larger number of experimental results is required to lower the level of reduction in the estimation of concrete contribution. Safety factors were not considered when calculating the ultimate shear strength of UHPC beams in the evaluation of the prediction model in this paper. For non-prestressed sections, the only two differences in $(V_{Rd,c})$ are that the value (0.24) is replaced by (0.18), and section height (h) is used instead of (z). Fibers contribution term is determined by quantifying the post-crack residual tensile strength resisting the main crack across the angle (θ) over (z) as shown in Fig. 1, Crane 2010 and Degen 2006. Fibers contribution term is calculated as follows:

$$V_{Rd,f} = \frac{A_{fv}\,6_{Rd,f}}{\tan(\theta)} \text{ [N]} \tag{15}$$

$$6_{Rd,f} = \frac{1}{K\,\gamma_{cf}} \frac{1}{w_{lim}} \int_0^{w_{lim}} 6_f\,(w).\,dw \text{ [MPa]} \tag{16}$$

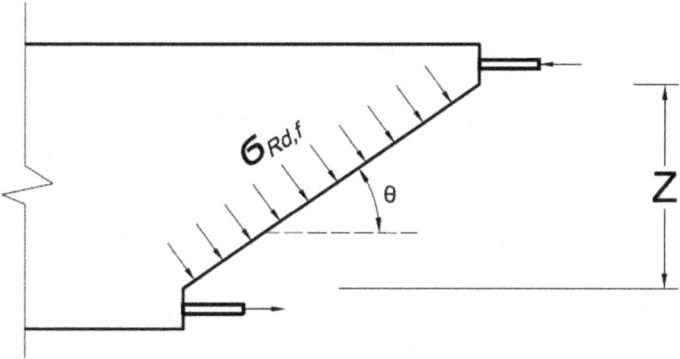

Fig. 1. Tensile stresses carried by steel fibers in a general beam section according to the Modified Compression Field Theory, Degen 2006

Where, (A_{fv}) is the area of fiber effect $(A_{fv} = b_w z)$ [mm^2]; $(\sigma_{Rd,f})$ is the residual tensile strength of the fiber-reinforced cross-section [MPa]; (θ) is the angle between the principal compression stress and the beam axis in degrees, a minimum value of 30 degrees is recommended. $(\sigma_{Rd,f})$ is estimated by the summation of the area under the stress-crack width curve of a three-point bending curve as shown in Eq. (14). (K) is a reduction factor used to consider the difference between fibers orientation of the prism and the actual orientation of the fibers in the future structure; the (K) factor typically ranges between 1.0 to 1.4. (w_{lim}) is the maximum of the ultimate crack width reached at the ultimate limit state bending moment of the section at the outer tension fiber, or the admissible crack width (recommended as 0.3 mm). Stress versus crack width $(\sigma_f(w))$ relation is determined by a bending test on UHPC prisms (six prisms are required). Three-point bending tests are performed on notched specimens, while four-point tests are used for un-notched specimens. The dimensions of the test prisms depend on the length of fibers (l_f). For $(l_f \leq 15$ mm (0.6 inch)), $70 \times 70 \times 280$ mm $(2.8 \times 2.8 \times 11$ in.) prisms are recommended, and for $((0.6$ in.$)$ 15 mm $< l_f \leq 20$ mm (0.8 in.)), $100 \times 100 \times 380$ mm $(4 \times 4 \times 15.6$ in.) prisms shall be used. The depth of the notch is equal to 10% of the prism height to enable an efficient localization of the crack while minimizing the risk of cracking outside the notch location. The distance between bearing points must be three times the depth of the prism. The residual tensile strength can be also estimated by means of direct tension tests on un-notched prisms.

3   Shear Experiments

Shear test data were collected from 15 research programs and focus on UHPC beams reinforced longitudinally with conventional reinforcement (f_y ranging between 400 to 600 MPa (58 to 87 ksi)) or prestressing strands (f_{pu} ranging between 1700 to 1860 MPa (246 to 270 ksi)). All UHPC mixes in this study have straight fibers with a tensile strength in the range of 1800 to 2600 MPa (260 to 377 ksi), except Voo et al. 2006, where a combination of straight and end hooked fibers were used in some tests. All the specimens considered in this study had a shear span to depth ratio of at least 2.3

and experienced a diagonal tension failure. Figure 2 shows the cross-section, width, and height of the considered fifteen experimental programs in this paper.

Hegger et al. 2004 performed an experiment on an I-beam reinforced with 8 strands on the bottom flange to investigate the bond anchorage behavior of the strands, in addition to the shear strength of the section. Flexural tensile strength of 40 MPa (5.8 ksi) was achieved for the 2.5% fiber volume mix having fiber length of 13 mm (0.5 in.). The UHPC beam showed a brittle shear failure mode similar to other specimens made with high-strength concrete. The degree of utilization of the prestressing force was approximately 80% at the time of failure.

Voo et al. 2006 performed seven shear tests on prestressed I-beams having 12 strands (15.2 mm diameter) on the bottom flange and 6 strands on the top flange. Top strands were tensioned to carry half the prestressing force of the bottom strands. In five beams, the bottom strands were tensioned to 15% of their yield strength (1750 MPa (254 ksi)), one beam was not tensioned at all, and one beam had the bottom strands tensioned to 30% of their yield strength. Bearing supports were placed 250 mm (10 in.) from the girder ends. Transfer length is reported to be about 250 mm (10 in.) for the 12.7 mm (0.5 in.) diameter strands and 356 mm (14 in.) for the 15.2 mm (0.6 in.) diameter strands Bertram et al. 2012, and Russell et al. 2013. It was noticed that the failure crack pattern extended horizontally at the transition between the flanges (with prestressing strands) and the web until reaching the supports. The total fiber volume fraction was 2.5%. However, the used steel fibers were a combination of straight (type I) and end-hooked (type II) types. Specimens SB1, SB2, and SB3 contained only type I fibers, while specimens SB4, SB5, SB7 contained 1.25%, 1%, 0.62% type II fibers respectively, and specimen SB6 contained only type II fibers. Measured crack angles ranged between 21 to 37°. Average flexural strength from notched three-point bending tests done on 100 × 100 mm (3.9 × 3.9 in.) prisms spanning 400 mm (15.7 in.) and notch depth of 25 mm (1.0 in.) was 23.5 MPa (3.4 ksi).

Graybeal 2006a, b performed three shear experiments on prestressed AASHTO Type II girders having 24 strands in the bottom flange (12 strands were debonded). Debonding effects of prestressing strands at the end of the girders were minimized by placing the bearing supports 4 feet from the girder ends (except girder 14S end bearing was placed 150 mm from girder end). A simplified prediction model for shear strength was proposed by calculating the diagonal tension carried by UHPC. The post-cracking diagonal tension strength for girders 24S and 14S was calculated inversely from the measured shear strength and resulted in 15.9, and 12.4 MPa (2.3, and 1.8 ksi) respectively. The same procedure was followed on small-scale tests and the post-cracking diagonal tension strength was estimated to be 9.0 MPa (1.3 ksi) which was considered a lower bound for un-reinforced webs with 2% steel fibers volume fraction.

Hegger and Bertram 2008 performed shear experiments on prestressed I-beams having 9 bottom strands (with 2 strands debonded). The testing was done to evaluate the bond anchorage of the strands and the shear strength of the section with and without web openings. Five specimens without web openings were only considered in this study. Crack angle ranged between 20 and 24° for the mixes having 0.9% fiber volume fraction.

Graybeal 2009 performed several flexural and shear tests on pi-girders to optimize the design of the section using UHPC. Strands were placed symmetrically in the two

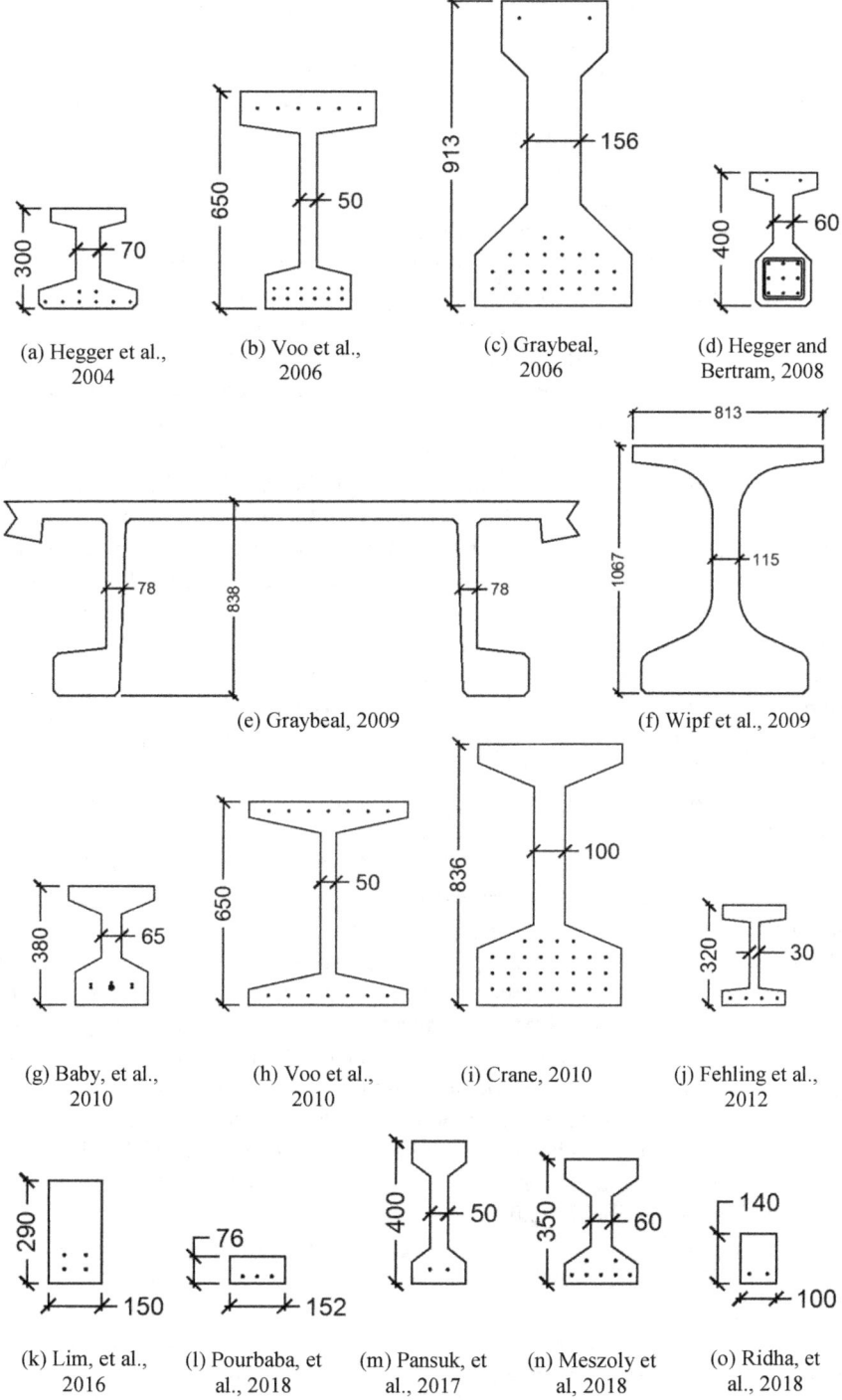

Fig. 2. Girder cross-sections of the considered fifteen experimental programs

bottom flanges, each flange had 11 strands with 3 strands debonded for 305 mm (12 in.), and 2 strands debonded for 1828 mm (72 in.). Two static shear experiments were performed with different shear spans on girders P2 and P4 that showed diagonal tension failure. For girder P2, supports were placed 610 mm (24 in.) away from the girder end to minimize the effect of debonding strands. While at girder P4 supports were placed only 152 mm (6 in.) from girder end. The prestressing force in the 1828 mm (72 in.) debonded strands was not considered when predicting the ultimate shear strength in this study as the debonding length is covering a significant portion of the shear span. Split cylinder cracking for the UHPC used in this program was 11.7 MPa (1.7 ksi), and direct tension cracking strength ranged between 9.7 to 11.0 MPa (1.4 to 1.6 ksi).

Wipf et al. 2009 performed a shear experiment on one large scale prestressed I-beam. The beam had a slightly modified Iowa DOT Bulb Tee C standard cross-section. The beam contained 47 strands in the bottom flange, 8 strands were debonded for 1067 mm (42 in.), and 16 strands were debonded for 1981 mm (78 in.), 5 strands were harped towards the top of the web at the girder ends. The prestressing force in the 1981 mm (78 in.) de-bonded strands were not considered when predicting the ultimate shear strength in this study as the debonding length is covering a significant portion of the shear span. The measured crack angle at failure was 25.3 degrees. Shear prediction analysis was performed and calibrated according to the test results based on the modified compression field theory and AFGC 2002. The fiber contribution factor is determined using the maximum tensile strength of UHPC which was recommended to be taken as 11.7 MPa (1.7 ksi) for the used mix having a 2% fiber volume fraction.

Baby et al. 2010 performed shear experiments on I-beams with the main test variables as the prestressing force (3 tests on prestressed, and 2 tests on non-prestressed beams) and fiber volume fraction (2 or 2.5%). Bearing supports were placed 500 mm (19.7 in.) from the beam end. Test results from this program were later compared to the prediction models of RILEM TC 162-TDF 2003, *fib* Model Code 2010, and AFGC 2002 by Baby et al. 2013a, b. It was concluded that RILEM TC 162-TDF 2003 and *fib* Model Code 2010 give similar predictions that are excessively conservative, while the AFGC 2002 provided a reasonable and conservative prediction.

Voo et al. 2010 performed shear experiments on prestressed I-beams having a slender web (50 mm (2 in.) thickness). The beams cross-section was symmetric about the neutral axis and had six strands on the top flange and six strands on the bottom flange having the same prestressing force. Bearing supports were placed 300 mm (11.8 in.) away from the girder end. Failure mode occurred from a tensile fracture across a single dominant crack or from a combination of cracks leading to the formation of a dominant crack. Average flexural strength from un-notched four-point bending tests done on 100 × 100 mm (3.9 × 3.9 in.) prisms spanning 300 mm (11.8 in.) was 14.0 MPa (2.0 ksi) for the mixes having 15 mm (0.6 in.) fiber length.

Crane 2010 performed six shear experiments on 835 mm (32 in.) deep bulb-tee girders with 200 mm (8 in.) thick cast in place high-performance concrete deck having 84 MPa (12.2 ksi) compressive strength. Crack angles ranged between 23 to 34°. Due to the large difference in concrete properties between girder and deck, the effective shear depth was based only on the girder and not the composite section. Measured shear capacities were compared against prediction models from two approaches. The first approach was based on calculating the direct tensile strength of the girder web over the failure angle, a direct tension strength of 9.7 MPa (1.4 ksi) was used based on previous research of a similar mix. The second approach was based on the AFGC 2002 in which a separate fiber contribution term is introduced, the residual rupture/tensile strength was taken as 6.9 MPa (1.0 ksi) based on previous research. It was concluded that the AFGC 2002 model provided a closer estimate to the measured shear strength than the direct tension approach with an average of less than 15% than the average measured strength.

Fehling and Thiemicke 2012 and performed shear experiments on rectangular non-prestressed beams having an I-shaped section on their shear span at which diagonal tension failure occurred, the remainder of the span 22 was a rectangular section. Bending tensile strength of 40×40 mm (1.6×1.6 in.) prisms having a length of 160 mm (6.3 in.) was 23.4 MPa (3.4 ksi) for the prisms having 1% fiber volume fraction. Crack angle ranged between 30 and 45°.

Lim and Hong 2016 performed shear experiments on rectangular beams with the main test variable as the transverse shear reinforcement ratio. Only one specimen (out of four) contained no transverse reinforcement and showed diagonal tension failure which was considered in this study. Shear strength was predicted using several research models, as well as the AFGC 2013 prediction model. All considered researcher prediction models underestimated the shear strength significantly, while the AFGC 2013 provided reasonable estimation (1.37 times the measured shear strength). Average direct tensile strength was 11.5 MPa (1.7 ksi) for the used UHPC mix which was used to predict the AFGC 2013 shear strength.

Pourbaba et al. 2018 and performed several shear experiments on non-prestressed rectangular beams with the main test variables as section dimensions, longitudinal reinforcement ratio, and shear span-to-depth ratio. Measured shear strength was compared against some code prediction models which were found to be excessively conservative.

Pansuk et al. 2017 performed four shear experiments on I-beams with main test variables as fiber volume fractions and transverse reinforcement ratio. The average splitting tensile strength was 16.5 and 15.2 MPa (2.4 and 2.2 ksi) for 1.6% and 0.8% fiber volume fraction mixes respectively. *fib* Model Code 2010 and AFGC 2002 prediction models were compared to measured shear strength. It was concluded that the *fib* Model Code 2010 model provided a much higher margin of safety compared to AFGC 2002.

Meszoly et al. 2018 performed shear experiments on I-beams with the main test variables as fiber volume fraction and transverse shear reinforcement. Measured shear strength was compared to the AFGC 2013 prediction model, ($\sigma_{R,f}$) was determined by performing flexural tests on prisms $150 \times 150 \times 700$ mm^3 ($5.9 \times 5.9 \times 27.6$ in^3). Results were back-calculated with an inverse analysis procedure as recommended and ($\sigma_{R,f}$) was estimated to be 5.0 and 6.3 MPa (0.73 and 0.91 ksi) for the mixes with 1% and 2% fiber volume fraction respectively.

Ridha et al. 2018 performed shear experiments on rectangular beams with the main test variables as the shear span-to-depth ratio, longitudinal reinforcement ratio, and fiber volume fraction. Average splitting tensile strength was 9.0, 11.0, 14.5, and 15.9 MPa (1.3, 1.6, 2.1, and 2.3 ksi) for mixes having 0.5%, 1.0%, 1.5%, and 2.0% fiber volume fraction, respectively.

4 Evaluation of Prediction Models

Steel fibers contribution is quantified using the post-cracking residual tensile strength of UHPC in all models. One assumption on the tensile strength of UHPC was made according to previous literature to reasonably estimate the fiber contribution. The German guidelines for UHPC DAfStb 2017 provides a conversion factor for the derivation of axial tensile strength from bending tests. A conversion factor of 0.37 is specified to be multiplied by the post-cracking flexural tensile strength to get the axial tensile strength. The Swiss standard SIA 2052 2016 provides a similar conversion factor of 0.383. This factor is based on having the neutral axis located at a distance of 0.82 times the prism height from the outermost tensioned fiber. In this section, the average value of these two factors (0.377) is multiplied by the maximum measured flexural stress to obtain the ultimate tensile strength.

Prediction models will be evaluated using the average measured-to-predicted shear strength of all collected data points. The upper limit of shear strength of conventionally reinforced concrete beams according to AASHTO LRFD 2017 and ACI 318 2014 is plotted as a reference horizontal line with the measured shear strength of UHPC beams. The limits are calculated according to Eqs. 17 and 18 for AASHTO 2017 and ACI 318 2014 respectively.

$$\frac{V_c + V_s}{b_w d} \leq 0.25 f_c' \quad [MPa] \tag{17}$$

$$\frac{V_c + V_s}{b_w d} \leq 1.08 \sqrt{f_c'} \quad [MPa] \tag{18}$$

Where, (V_c) is the concrete contribution to shear strength (N); (V_s) is the transverse reinforcement contribution to shear strength (N); (f_c') is the cylinder compressive strength in (MPa); (b_v) is the minimum web width (mm); (d_v) is the effective shear depth between the resultants of the tensile and compressive forces due to flexure (mm).

Girder compressive strength was considered as 69 MPa (10 ksi) as an upper limit for the conventional concrete compressive strength as required by AASHTO LRFD 2017 and ACI 318 2014. Measured shear strengths of UHPC girders without transverse reinforcement exceeded the ACI 318 2014 upper limit for most of the data points, and can meet and exceed the upper limit of AASHTO 2017 as shown in Figs. 3, 4 and 5.

4.1 RILEM TC 162-TDF 2003

Table 1 presents a review of some of the three-point bending tests done to quantify the residual tensile strength of UHPC. The 2% fiber volume fraction mix is the most commonly used and tested mix. Averaging the available results to find a reasonable assumption for that fiber volume fraction mixes for $(f_{R,4})$ would be 29.6 MPa (4.3 ksi). This assumption corresponds well to having the direct tensile strength of 11.0 MPa (1.6 ksi) (reported by Graybeal 2006a, b, Wipf et al. 2009, and Lim and Hong 2016) divided by the conversion factors discussed above. For the 2.5% and 1% mixes, a reasonable assumption according to Table 1 for $(f_{R,4})$ would be to use the 40.0 and 2.8 MPa (5.4 and 0.4 ksi) respectively. Mixes having different fiber volume fractions (like 1.5%) are

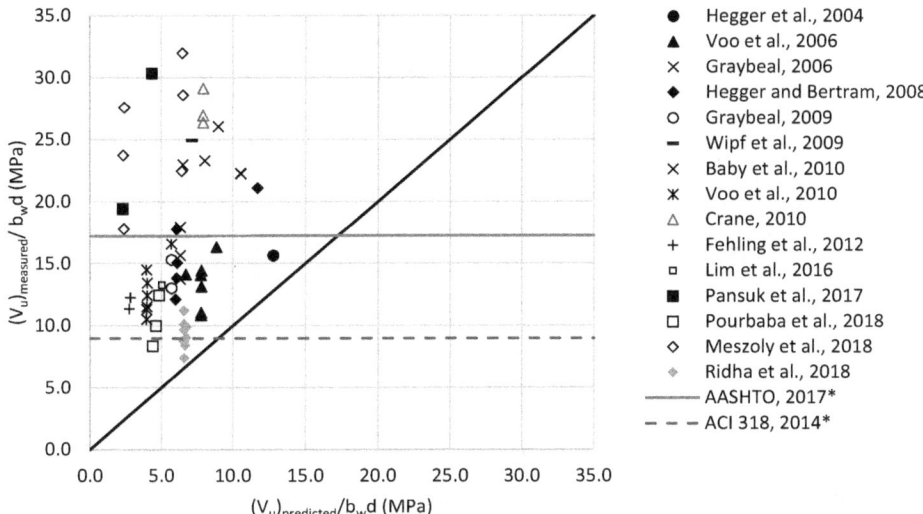

Fig. 3. Measured versus predicted shear strength according to RILEM TC 162-TDF, 2003 (note: 1 MPa = 0.145 ksi) *Upper limits of the combined contribution of concrete and transverse reinforcement of conventionally reinforced concrete (69 MPa)

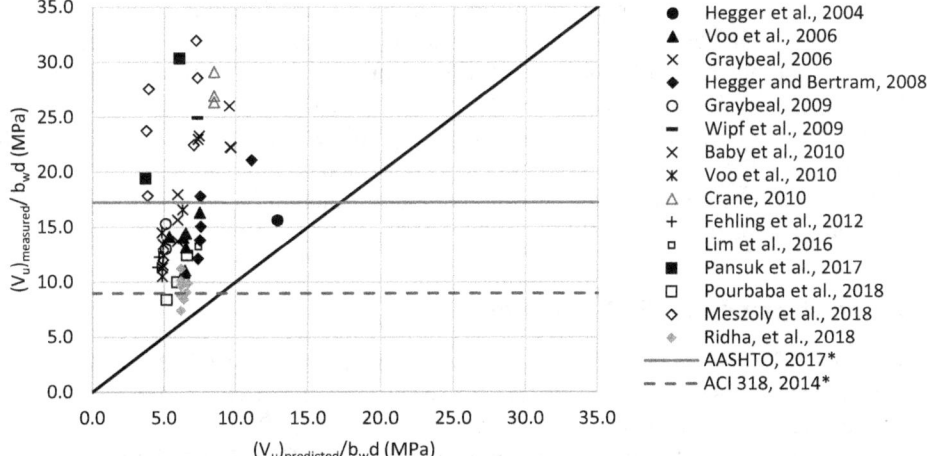

Fig. 4. Measured versus predicted shear strength according to the *fib* Model Code 2010 (note: 1 MPa = 0.145 ksi) *Upper limits of the combined contribution of concrete and transverse reinforcement of conventionally reinforced concrete (69 MPa)

interpolated according to these values. Figure 3 shows the measured versus predicted shear capacities plot for the considered data points. The average measured-to-predicted shear strength was 3.1, with a standard deviation of 2.1. It can be noticed that safety margin increases with the increase of the measured shear strength.

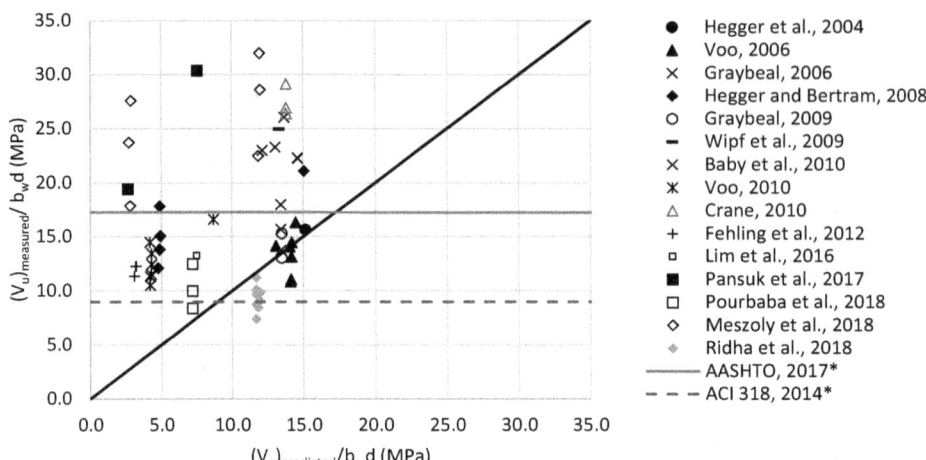

Fig. 5. Measured versus predicted shear strength according to French Standard NF P 18-710 2016 (note: 1 MPa = 0.145 ksi) *Upper limits of the combined contribution of concrete and transverse reinforcement of conventionally reinforced concrete (69 MPa)

4.2 *fib* Model Code 2010

A reasonable assumption for (f_{ftuk}) for the 2% fiber volume fraction mix is 11.0 MPa (1.6 ksi) based on Eqs. (10) and (11), and having the flexural tensile strengths $(f_{R,i})$ according to Table 1. UHPC with different fiber fractions were interpolated according to Table 1. Figure 4 shows the measured versus predicted shear capacities plot for the considered data points. The average measured-to-predicted shear strength was 2.6, with a standard deviation of 1.2. It can be noticed that the standard deviation is significantly lower than the other considered international codes indicating a higher consistency in the prediction model.

4.3 French Standard, NF P 18-710 2016

Graybeal 2019 estimated the expected range for direct tension sustained post-cracking tensile strength between 5.5 to 8.3 MPa (0.8 to 1.2 ksi). Haber et al. 2018 reported the direct tensile strength as 8.4 MPa (1.2 ksi) for UHPC with 2% fiber volume fraction at 28-days. The direct tensile strength corresponds to the French Standard 2016 post-cracking residual flexural tensile strength $(\sigma_{Rd,f})$. The assumed value of $(\sigma_{Rd,f})$ for the 2% fiber volume fraction UHPC was selected conservatively as 6.9 MPa (1.0 ksi). Data points from two research programs were overestimated: Voo et al. 2006 where the over estimation can be attributed to the relatively slender 50 mm (2 in.) web thickness, Ridha et al. 2018 where the over estimation can be attributed to the used UHPC mix which had a lower compressive strength than what is typically achieved by Ductal. Also, Specimen P4-57SH data point from Graybeal 2009 was slightly overestimated and the measured-to-predicted shear strength was 0.97) which can be attributed to the increased number of debonded strands and the relatively slender web thickness 84 mm (3.3 in.) compared the bottom flanges. Figure 5 shows the measured versus predicted shear strength plot for

Table 1. Three-point bending tests on notched prisms to evaluate residual tensile strength

Reference	Fiber volume fraction (V_f)	Fiber length (l_f) (mm)	Fiber diameter (Φ_f) (mm)	Cylinder compressive strength (f'_c) (MPa) (ksi)	Prism cross section (b × h) (mm²)	Notch height (mm) (in.)	Span (mm) (in.)	Residual flexural tensile strengths ($f_{R,i}$) (MPa) (ksi)			
								$f_{R,1}$	$f_{R,2}$	$f_{R,3}$	$f_{R,4}$
Prem et al. 2012 (R1 Mix)	2.5%	13	0.15	180.0	70 × 70	21.0	300	45.5	49.6	42.7	40.0
Prem et al. 2012 (R2 Mix)	2%	13	0.15	170.0	70 × 70	21.0	300	37.2	40.0	37.2	34.5
Yang et al. 2010 (average)	2%	13	0.2	190.9	100 × 100	10.0	300	26.9	30.3	27.6	24.8
Graybeal, 2006a, b (M2P02)	2%	13	0.2	126.2	50 × 100	25.4	406	22.1	20.7	–	–
Zagon et al. 2016 (average)	1%	10	0.18	141.3	100 × 100	27.0	400	11.7	6.2	4.1	2.8

the considered data points. The average measured-to-predicted shear strength was 2.3, with a standard deviation of 1.8.

Figure 6 shows a comparison of the prediction models against the measured shear strength for the average of each experimental program. A summary of the evaluation of prediction models is presented in Table 2 with the main assumption taken for quantifying the post-crack residual tensile strength of the fiber contribution term. It can be noticed that the French Standard NF P 18-710 2016 provided the closest estimation of the shear strength followed by the *fib* Model Code 2010 model. However, the *fib* Model Code 2010 model had the least standard deviation which indicates providing the highest consistence in the shear strength prediction.

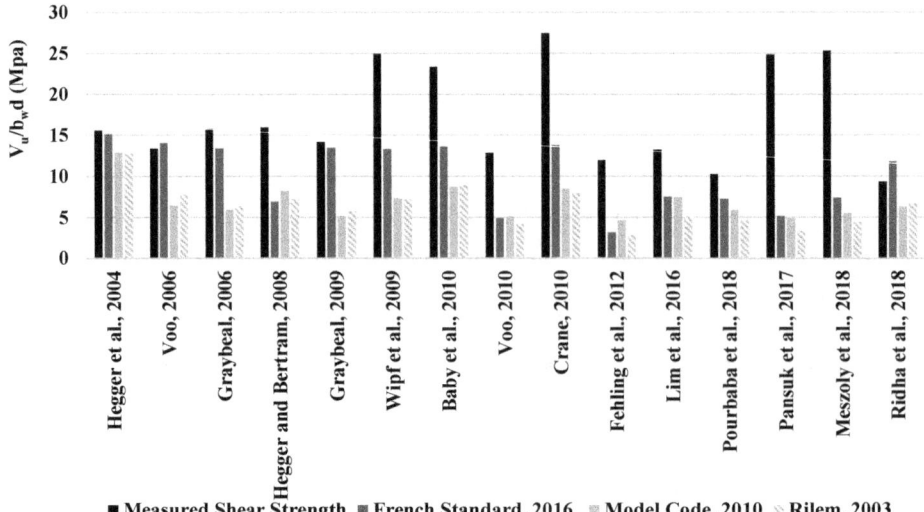

Fig. 6. Measured versus predicted shear strength according to the considered models for the average of each experimental program (note: 1 MPa = 0.145 ksi)

Table 2. Summary of evaluation of prediction models

International practice	Average measured-to-predicted shear strength $\frac{(V_u)_{measured}}{(V_u)_{predicted}}$	Standard deviation	Assumption for $V_f = 2\%$
RILEM TC 162-TDF (2003)	3.1	2.1	$f_{R,4} = 29.6$ MPa
fib Model Code 2010	2.6	1.2	$f_{ftuk} = 11.0$ MPa
French Standard NF P 18-710 2016	2.3	1.8	$\mathbf{6}_{Rd,f} = 6.9$ MPa

5 Conclusions

1. The considered prediction models underestimate the shear strength of UHPC beams.
2. The French Standard NF P 18-710 2016 model provided the closest prediction to the measured shear strength of UHPC beams.
3. The *fib* Model Code, 2010 model provided the highest consistency in the prediction of shear strength.
4. The RILEM TC 162-TDF 2003 model provided the least shear strength, and the highest scatter in its prediction.

References

AASHTO LRFD: Bridge design specifications. American Association of State Highway and Transportation Officials, Washington, D.C., eighth edn (2017)

ACI Committee 318: Building Code Requirements for Structural Concrete (ACI 318 14) and Commentary on Building Code Requirements for Structural Concrete (ACI 318R-14) (2014)

AFGC (Association Francaise de Génie Civil): Ultra high performance fibre-reinforced concretes. Interim Recommendations. AFGC publication, France (2002)

AFGC (Association Francaise de Génie Civil): Ultra high performance fibre-reinforced concretes, recommendations. AFGC publication, France (2013)

Baby, F., Marchand, P., Toutlemonde, F., Billo, J., Simon, A.: Shear resistance of ultra high performance fibre-reinforced concrete I-beams. In: FraMCoS7, pp. 1411–1417 (2010)

Baby, F., Marchand, P., Toutlemonde, F.: Shear behavior of ultrahigh performance fiber-reinforced concrete beams. I: experimental investigation. J. Struct. Eng. **140**(5), 04013111 (2013a)

Baby, F., Marchand, P., Toutlemonde, F.: Shear behavior of ultrahigh performance fiber-reinforced concrete beams. II: analysis and design provisions. J. Struct. Eng. **140**(5), 04013112 (2013b)

Bertram, G., Hegger, J.: Bond behavior of strands in UHPC–tests and design. In: Proceedings of the 3rd International Symposium on UHPC, Kassel, Germany, pp. 525–532 (2012)

Binard, J.P.: UHPC: a game-changing material for PCI bridge producers. PCI J. **62**, 34–36 (2017). https://doi.org/10.15554/pcij62.2-01

Crane, C.K.: Shear and shear friction of ultra-high performance concrete bridge girders. Dissertation, Georgia Institute of Technology (2010)

DAfStb-Guideline: German Committee for Structural Concrete, Ultra-High Performance Concrete (Draft), Berlin (2017)

Degen, B.E.: Shear design and behavior of ultra-high performance concrete (2006)

European pre-standard: ENV 1992-1-1: Eurocode 2: Design of Concrete Structures – Part 1: General rules and rules for buildings (1991). https://standards.iteh.ai/catalog/standards/cen/be5ea153-9eb7-4d19-b90b-5c3d40412a95/env-1992-1-1-1991

EN, BS.: "14651." Test Method for Metallic Fibre Concrete-Measuring the Flexural Tensile Strength (Limit of Propportionally (LOP), Residual) (2007), pp. 1–20

EPFL-Swiss Federal Institute of Technology: Standard: Ultra-High Performance Fibre Reinforced Cement-based composites (UHPFRC) Construction material, dimensioning and application, SIA 2052 (2016)

Fehling, E., Thiemicke, J.: Experimental investigations on I-Shaped UHPC-Beams with combined reinforcement under shear load. Ultra high performance concrete and nanotechnology in construction. In: 3rd International Symposium on Ultra High Performance Concrete and Nanotechnology for High Performance Construction Materials. Kassel University Press GmbH, Kassel (2012)

French standard, NF P 18-710: National addition to Eurocode 2 — Design of concrete structures: specific rules for Ultra-High Performance Fibre-Reinforced Concrete (UHPFRC) (2016)

Foster, S.J., Agarwal, A., Amin, A.: Design of steel fiber reinforced concrete beams for shear using inverse analysis for determination of residual tensile strength. Struct. Concr. **19**(1), 129–140 (2018)

Graybeal, B.A.: Structural Behavior of a Prototype Ultra-High Performance Concrete Pi-Girder. No. FHWA-HRT-10-027. United States. Federal Highway Administration. Office of Infrastructure Research and Development (2009)

Graybeal, B.A.: Design and Construction of Field-Cast UHPC Connections, FHWA, U.S. Department of Transportation, Publication No: FHWA-HRT-19-011 (2019)

Graybeal, B.A.: Material property characterization of ultra-high performance concrete. Publication No. FHWA-HRT-06-103. United States. Federal Highway Administration. Office of Infrastructure Research and Development (2006a)

Graybeal, B.A.: Structural Behavior of Ultra-High Performance Concrete Prestressed I-Girders. Publication No. FHWA-HRT-06-115. United States. Federal Highway Administration. Office of Infrastructure Research and Development (2006b)

Haber, Z.B., De la Varga, I., Graybeal, B.A., Nakashoji, B., El-Helou, R.: Properties and behavior of UHPC-class materials (No. FHWA-HRT-18-036). United States. Federal Highway Administration. Office of Infrastructure Research and Development (2018)

Hegger, J., Tuchlinski, D., Kommer, B.: Bond anchorage behavior and shear capacity of ultra high performance concrete beams. In: Proceedings of the International Symposium on Ultra High Performance Concrete (2004)

Hegger, J., Bertram, G.:. Shear carrying capacity of ultra-high performance concrete beams. In: Proceedings of the International fib Symposium 2008 - Tailor Made Concrete Structures: New Solutions for Our Society, 96 (2008)

International Federation for Structural Concrete CEB-fib: CEB-FIP Model Code, Final Draft, vol. 1. Lausanne, Switzerland (2010)

Lim, W.-Y., Hong, S.-G.: Shear tests for ultra-high performance fiber reinforced concrete (UHPFRC) beams with shear reinforcement. Int. J. Concr. Struct. Mater. **10**(2), 177–188 (2016). https://doi.org/10.1007/s40069-016-0145-8

Mészöly, T., Randl, N.: Shear behavior of fiber-reinforced ultra-high performance concrete beams. Eng. Struct. **168**(April), 119–127 (2018). https://doi.org/10.1016/j.engstruct.2018.04.075

Pansuk, W., Nguyen, T.N., Sato, Y., Den Uijl, J.A., Walraven, J.C.: Shear capacity of high performance fiber reinforced concrete I-beams. Constr. Build. Mater. **157**(December), 182–193 (2017). https://doi.org/10.1016/j.conbuildmat.2017.09.057

Prem, P.R., Bharatkumar, B.H., Iyer, N.R.: Mechanical properties of ultra high performance concrete. World Acad. Sci. Eng. Tech. **68**, 1969–1978 (2012)

Pourbaba, M., Joghataie, A., Mirmiran, A.: Shear behavior of ultra-high performance concrete. Constr. Build. Mater. **183**, 554–564 (2018)

Ridha, M.M., Al-Shaarbaf, I.A., Sarsam, K.F.: Experimental study on shear resistance of reactive powder concrete beams without stirrups. Mech. Adv. Mater. Struct. **27**, 1–13 (2018)

Russell, H.G., Graybeal, B.A., Russell, H.G.: Ultra-high performance concrete: A state-of-the-art report for the bridge community (No. FHWA-HRT-13-060). United States. Federal Highway Administration. Office of Infrastructure Research and Development (2013)

Vandewalle, L.: Design method for steel fiber reinforced concrete proposed by RILEM TC 162-TDF. In: Fifth International RILEM Symposium on Fibre-Reinforced Concrete (FRC), pp. 51–64. RILEM Publications SARL (2000a)

Voo, Y.L., Foster, S.J., Gilbert, R.I.: Shear strength of fiber reinforced reactive powder concrete prestressed girders without stirrups. J. Adv. Concr. Tech. **4**(1), 123–132 (2006)

Voo, Y.L., Poon, W.K., Foster, S.J.: Shear strength of steel fiber-reinforced ultrahigh-performance concrete beams without stirrups. J. Struct. Eng. **136**(11), 1393–1400 (2010)

Vandewalle, L.: Recommendations of RILEM TC 162-TDF: test and design methods for steel fibre reinforced concrete. Mater. Struct./Materiaux et Constructions **33**(225), 3–5 (2000b)

Wipf, T., Phares, B., Sritharan, S., Degen, B, Giesmann, M.T.: Design and Evaluation of a Single-Span Bridge Using Ultra-High Performance Concrete (2009)

Yang, I.H., Joh, C., Kim, B.-S.: Structural behavior of ultra high performance concrete beams subjected to bending. Eng. Struct. **32**(11), 3478–3487 (2010)

Zagon, R., Matthys, S., Kiss, Z.: Shear behaviour of SFR-UHPC I-shaped beams. Constr. Build. Mater. **124**, 258–268 (2016)

Evaluation of the Performance of Gravel Road with Base Course Reinforced with Do-Nou

Yoshinori Fukubayashi[1]([✉]), Sohei Sato[2], Atsushi Koyama[1], and Daisuke Suetsugu[1]

[1] University of Miyazaki, Miyazaki, Japan
fukubayashi@cc.miyazaki-u.ac.jp
[2] Graduate School of University of Miyazaki, Miyazaki, Japan

Abstract. In developing countries, improvement of trafficability of rural roads, which are generally earth or graveled, are crucial for rural development. Under the circumstances with financial constraints of those countries, locally available material-based approach towards improvement of trafficability are regarded as one of the practical measures. Do-nou, which is Japanese term for soilbag, have been utilized for reinforcing base course. Do-nou method is requiring only labor even for compaction of base material and used polyester fiber woven bags for crops, fertilizer, etc. as one of geotextile. In 29 countries of Asia, Africa and the Pacific, about 180 km rural road has been improved using the Do-nou method, by the NGO which one of the authors is belong to. Because of the easiness and effectiveness of Do-nou method, it enables minor and small enterprises, generally operated by the youth, to be involved in road construction implemented by public works. In order to improve trafficability of rural roads while solving social problem, high unemployment rate among the youth in developing countries, the method needs to be adopted by road authorities for further extension. In this study, therefore, the performance of gravel road with base course consist of Do-nou have been evaluated through the series of full-size driving tests. The settlement of road surfaces with conventionally designed base course and with that consist of Do-nou with the same thickness subjected to the traffic load were measured and compared. Dynamic cone penetration index of these base course structure was also compared. Load distribution effects were also examined using earth pressure gages located on subgrade and road surface during driving tests. The mechanism of reducing settlement of road surface with Do-nou base course were verified. The performance of base course with Do-nou are defined in comparison to conventional base course structure, which enable road authorities adopt Do-nou method.

1   Introduction

In developing countries, road pavement rate with asphalt and concrete remain low compared to that in developed countries due to the financial constraints. It means that the road surface in rural area is mostly, earth, engineered natural or gravel. It is true that the traffic volume passing on rural roads is low, but at the same time, it is crucial for rural development that those roads keep all-weather access.

With this regard, the low volume road design guidelines have been established in many developing countries aiming to expand all-weather access road network in their

H. Shehata and S. El-Badawy (Eds.): *Sustainable Issues in Infrastructure Engineering*, SUCI, pp. 40–54, 2021.
https://doi.org/10.1007/978-3-030-62586-3_4

entire countries. Since 1980s, the International Labor Organization (ILO) have accumulated a lot of experience of implementing rural road rehabilitation and maintenance projects in developing countries over the world by applying the labor-based technology (LBT) (Donnges et al. 2007). LBT is defined as the construction technology which, while maintaining cost competitiveness and acceptable engineering quality standards, maximizes opportunities for the employment of labor (skilled and unskilled) together with the support of light equipment and with the utilization of locally available materials and other resources (Johannessen 2008). LBT has been adopted in most of the low volume road design guidelines and also demonstrated mostly with financial and technical support from donor agencies (Photo 1).

Photo 1. Construction of gravel road with LBT, compaction with equipment

Though tremendous projects have been implemented to improve the rural road networks using LBT in developing countries, generally, more than 50% of rural roads are fallen into poor conditions, especially in low or low middle-income countries (Donnges et al. 2007). The rehabilitated roads and newly constructed roads have been deteriorated into unmaintainable conditions due to no adequate maintenance. This phenomenon has existed for more than 30 years.

Aiming to change the situation where rural people have been suffering from poor access to social services and markets, Fukubayashi and Kimura (2014) have developed the innovative method for road improvement, which was locally available material based and labor intensive to enable and encourage road side communities to conduct the civil works for maintaining and improving trafficability of the road networks by themselves. The method was a spot improvement using Do-nou, which is a Japanese term for soil bag. Do-nou method is requiring only labor for filling soil, laying and even compaction and new or second-hand polyester fiber woven bags for crops, fertilizer, etc. as kind of geotextile. Do-nou filled with soil material are laid on subgrade with manual compaction up to required number of layers to build base course, which bear traffic load (Photo 2). Do-nou own high bearing capacity through reinforcing soil filled in the bags with the tensile strength generated in the wrapping bags during compaction (Matsuoka and Liu 2006).

Photo 2. Construction of base course with Do-nou method

In 29 countries of Asia, Africa and the Pacific, about 180 km rural road has been improved using the Do-nou method (Fukubayashi and Kimura 2017). Empirically, the quality of the improved road with Do-nou method has been found to be satisfactory and reasonable considering the low cost and non-requirement of even light equipment, like compactors. Because of the easiness and effectiveness of Do-nou method, it enabled minor and small enterprises, generally operated by the youth, to be involved in road construction projects executed by road authorities in Kenya. In order to improve trafficability of rural roads together with solving social problem, high unemployment rate among the youth in developing countries, Do-nou method needs to be widely utilized with proper understanding of the performance of the road with Do-nou base course.

In order to evaluate the performance of gravel road with base course reinforced with Do-nou in comparison to those with conventional base course subjected to light equipment vibrating compaction, the series of full-size driving tests have been conducted. The settlement of road surfaces with conventionally designed base course and with that consist of Do-nou with the same thickness subjected to the same traffic load were measured and compared. The performance of base course in dry and wet conditions were examined considering road conditions in dry and rainy seasons. Based on the results of the measured earth pressure underneath the base course, the load distribution effects were also analyzed (Photo 3).

2 Full Size Driving Test

The series of full-size driving tests have been conducted at the free space in Field Science Center of University of Miyazaki in Japan (Photo 4). The lot with 3 m in width and 5 m in length was prepared through vegetation control and leveling. The in-situ soil characteristics are shown in Table 1. The original ground surface was modelled as subgrade for driving tests. Whenever starting a new case of driving test, before building base course, the water contents of subgrade surface were adjusted to be the natural water content initially measured. Portable cone penetration tests (JGS 1431) on subgrade were carried out before building the base course in order to confirm the bearing capacity of the subgrade were not varied through all the cases of experiments. As shown in Fig. 1, the cone penetration resistance measured 1.0 m below the subgrade surface was less than 1,000 kN/m^2, which meant the subgrade didn't own enough bearing capacity against traffic loads.

Photo 3. Installation of earth pressure gauges

Photo 4. Full-size driving test

Table 1. Characteristics of the subgrade soil

Soil classification	Volcanic cohesive soil
Density ρ_s (g/cm^3)	2.765
Natural water content ω_n (%)	52.6
Liquid limit ω_P (%)	82.8
Plastic limit ω_L (%)	43.8
Plasticity index I_P	38.9
Design CBR (%)	4.0

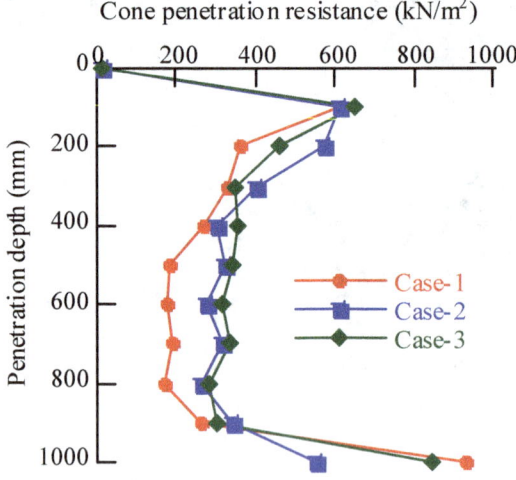

Fig 1. Cone penetration resistance of the measurement depth of subgrade

2.1 Design of the Base Course Thickness

In this study, the Design manual for low volume roads in Ethiopia (Ethiopian Road Authority 2011) was referred to design the base course structure for conventional method. The design catalog describing the thickness of base course stipulated in the design manual was chosen based on bearing capacity of subgrade and base course material. As shown in Table 1, the subgrade soil was volcanic cohesive soil with the designed California Bearing Ratio (CBR) 4%. The mechanically stabilized crushed stones produced in the quarry nearby the test field were utilized as base course material and filling material into Do-nou bags, considering the availability and consistence of the quality. The characteristics is shown in Table 2. In all the cases, the water content was adjusted to be close to the optimum water content. In Fig. 2, the grain size distribution curves of the subgrade soil and the mechanically stabilized crushed stones are shown. It was confirmed that the grain size distribution curves of the mechanically stabilized crushed stones were within the range of the base material standardized in the design manual and the other requirement, such as plasticity index were also complied to the standard set in the design manual. However, it should be noted that the modified CBR value, 101%, was much higher than that of base course material for gravel roads, which was 15–45%. In this study, to examine the differences of the performance of base course with Do-nou confinement effect with manual compaction and that built conventionally with compaction by vibrating compactor, the mechanically stabilized crushed stones were still used.

Table 2. Characteristics of base course material

Name	Mechanically stabilized crushed stones
Modified CBR (%)	101
Maximum dry density ρ_{dmax} (g/cm^3)	2.216
Optimum water content ω_{opt} (%)	5.5

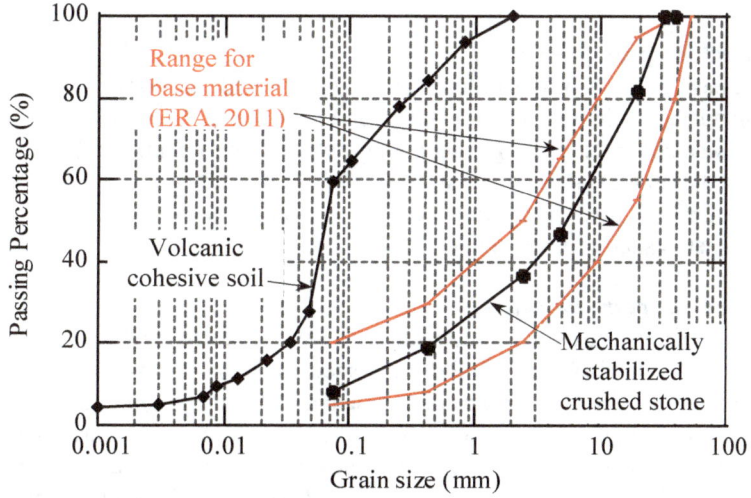

Fig 2. Grain size distributions of subgrade soil and base course material

2-ton dump truck was used for driving tests because it was one of the largest traffic loads in rural areas.

Considering the traffic in rural areas of developing countries, cumulative traffic load for the designed period 10 years was calculated with the assumption that the average daily vehicle number in one direction was 150 of 2-ton truck with 5% annual growth rate (ERA 2011). Thus, the traffic class was decided.

Finally, according to the design catalog, based on the subgrade strength class, types of base course material, and traffic classes, the base course thickness was decided as 225 mm.

2.2   Cases of Full-Size Driving Test

Driving tests were performed in the three cases as shown in Fig. 3. In all the cases, the thickness of base course was 225 mm, while the structures and compaction method was varied.

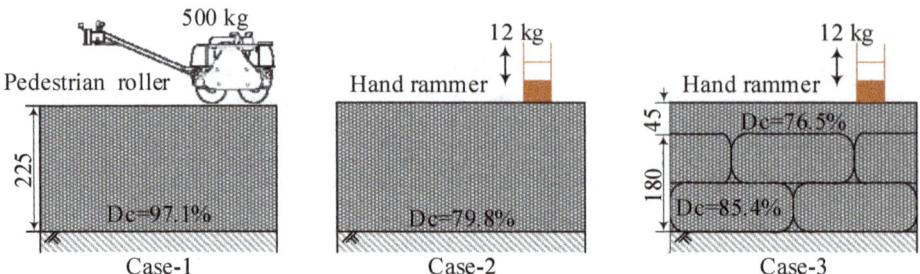

Fig 3. Cases of full-size driving tests

Base Course Structure for Case-1

In Case-1, base course was constructed in accordance with the conventional design and method standardized in the Design manual. The base was divided into three layers, and each layer was compacted using a pedestrian roller weighing about 500 kg. The roller passed over the same sections six times. It was found that the compaction degree reached to 97.1% which was measured through the test method for measuring in-situ soil density using compacted sand replacement (JGS 1611, 2003).

Base Course Structure for Case-2

The structure of the base course in Case-2 was consist of mechanically stabilized crushed stones as same as that in Case-1, while compacted manually. It was considered that even light equipment for compaction was often not available when road improvement works was initiated through self-reliance initiatives of communities in rural area of developing countries. The total compaction energy was governed to be same as Case-3. The base course was divided into three layers similarly as Case-1 and 3, and then, at each layer, a 400 mm × 400 mm square area was compacted manually with 20 times of hits of the hand rammer weighing 12 kg. Then, the adjacent square area was compacted and this was repeated until the entire area of the layer was compacted followed by the spread of base course material for the next layer. The compaction degree was 79.8% measured by the method for measuring in-situ soil density using core cutter (JGS 1613, 2003).

Base Course Structure for Case-3

In Case-3, the base course consisted of two layers of Do-nou and a layer of mechanically stabilized crushed stones with 45 mm thickness, which was fixed to make the total thickness of base course in Case-3 to be 225 mm as shown in Fig. 3. The dimensions and tensile strength of Do-nou bags utilized in this study are shown in Table 3, which are similar to the polyester fiber woven bags for crops, fertilizer, etc. available in developing countries. Each layer of Do-nou consisted of 60 pieces. After the compaction, the Do-nou filled with a designated volume of base course material became 40 cm in depth and length and 9 cm in height. Even after the compaction, the space between a Do-nou and the adjacent one was remained, which was filled with base course material and compacted manually. The next layer of Do-nou was laid alternately to the previous layer, then compacted. The surface layer on the second layer of Do-nou consisting of only base course material was compacted as same as Case-2. To calculate the compaction degree of filled material inside Do-nou bag after the compaction, the volume of Do-nou was obtained with the measured height of Do-nou and surface area measured from the image analysis. As a result, the compaction degree of base course wrapped with Do-nou bags was 85.4%. If the compaction degrees compared to that in Case-2, it can be said that with confinement effect from the wrapping Do-nou bags, compaction degree of base course material was increased.

Table 3. Characteristics of Do-nou bags

Material	Polyethylene			
Size (cm)	0.48 × 0.62			
Weight (g)	58			
Tensile Strength (N/5 cm)	Vertical direction	550	Horizontal direction	500
Tensile elongation (%)	Vertical direction	22	Horizontal direction	17

2.3   Estimation of Bearing Capacity of Base Course in Case-1–3

The bearing capacity of the base course in each case was evaluated by Dynamic cone penetration test conducted just after building the base course and at the tire pass after the driving tests. The DCP test is used as a test that can easily measure the bearing capacity of base course in developing countries. Before driving test, the CBR values converted from the measured dynamic cone index with the prescribed formula (Paige and Plessis 2009) were 14 to 28% for Case-1, and less than 10% for Case-2 and 3 as shown in Fig. 4.

2.4   Installation of Earth Pressure Gauge

Two earth pressure gauge (KDG-500 kPa) were installed on the subgrade in order to grasp and analyze the load distribution effects of those structures of base course (Photo 3). As shown in Fig. 5, one (Gauge A) was placed directly below the tire pass and the other (Gauge B) was placed 20 cm laterally away from Gauge A. In order to understand the tendency of the measured values under wheel load, an earth pressure gauge was installed on the asphalt pavement, and a 2-ton truck used for driving tests were run on it. The measured value was almost in agreement with the tire air pressure of 260 kPa same as wheel load estimated from the vehicle weight. When installed on the subgrade, it is difficult to measure the accurate stress due to the deformation of subgrade surface. However, the measured results can be utilized to analyze the tendency of load distribution effects of base course structures in all the cases.

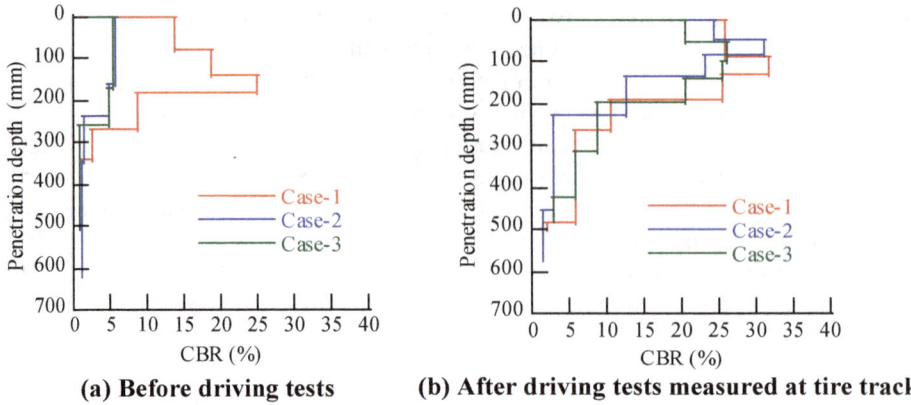

(a) Before driving tests   (b) After driving tests measured at tire track

Fig 4. Results of dynamic cone penetration tests

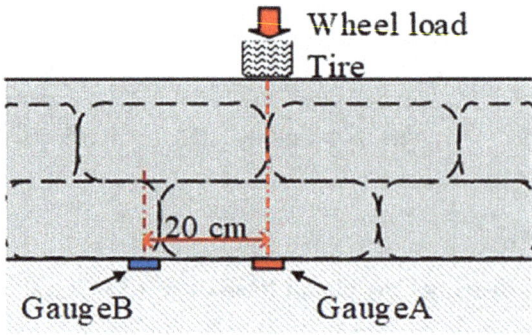

Fig 5. Earth pressure gauge installation position

2.5 Number of Passes Made by a 2-Ton Dump Truck and Measurement Items

After constructing the base course, the 2-ton dump truck drove forward and backward, at the speed of 5 km/h, over the same channel on the constructed base course (Photo 4). The 2-ton dump truck made 150 round trips (300 passes). After that, a volume of water equivalent to the average daily rainfall (during the rainy season in Ethiopia) was sprinkled over the entire area of the base course, then another 150 round trips were made. This was reflected with the scenario that road construction was conducted during the dry season, then the constructed roads were exposed to rainfall during the rainy season.

In all the cases, the settlement of the base course surface and earth pressure were measured at the designated number of passes. After the driving tests, the base course material was removed, and the settlement of subgrade was measured. By comparing the settlement of base course surface and subgrade measured before and after the driving test, the deformation of the base course was obtained.

3 Experimental Results and Discussion

3.1 Settlement of Base Course Surface

As the 2-ton dump truck passed on the base course, the surface near the tire track was settled and ruts were formed. Figure 6 shows the cross-section view of base course at the end of the driving tests, after 600 passes. The relationship between the settlement at the deepest point of the rut and the number of passes (maximum of 600 passes) for all the cases are shown in Fig. 7.

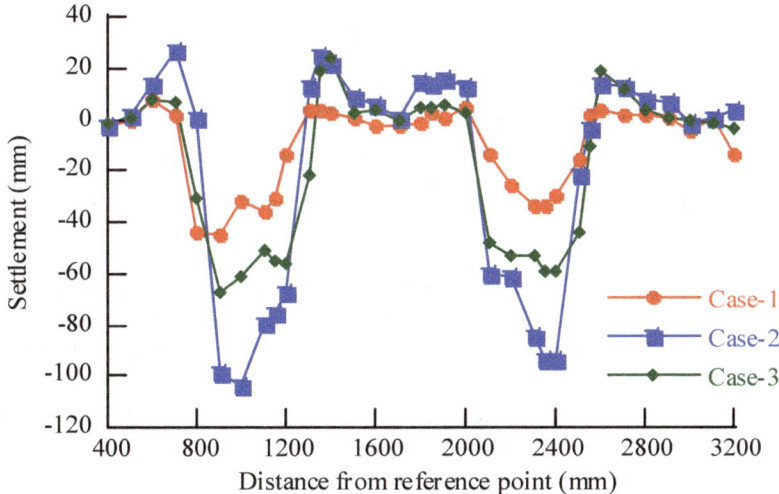

Fig 6. Cross-section view of base course after final pass

Settlement of Baes Course in Dry Condition

In all the cases, while the base course kept dry condition, the settlement didn't increase much after 100 passes. Case-1, which has the highest compaction degree, had the minimum settlement compared to those of Case-2 and Case-3. The differences of the settlement in Case-2 and Case-3 were not significant, while the compaction degree of base course in Case-3 was larger than that of Case-2. Actually, the settlement of Case-2 was smaller than Case-3. It can be considered that the spaces between Do-nou and the adjacent Do-nou were not compacted well and caused the larger settlement. It was one of the advantages that Do-nou could be handled and compacted manually because of the volume and weight of each Do-nou, while the application of Do-nou to base course caused noncontinuous characteristics of base course layer. It was also possible that the surface layer on the second Do-nou layer was thin and less compacted so that the material was easily moved aside under wheel load resulting in the larger settlement of the surface.

Settlement of Baes Course in Wet Condition

After sprinkling water, the minimum settlement was observed in Case-3, and the largest

50   Y. Fukubayashi et al.

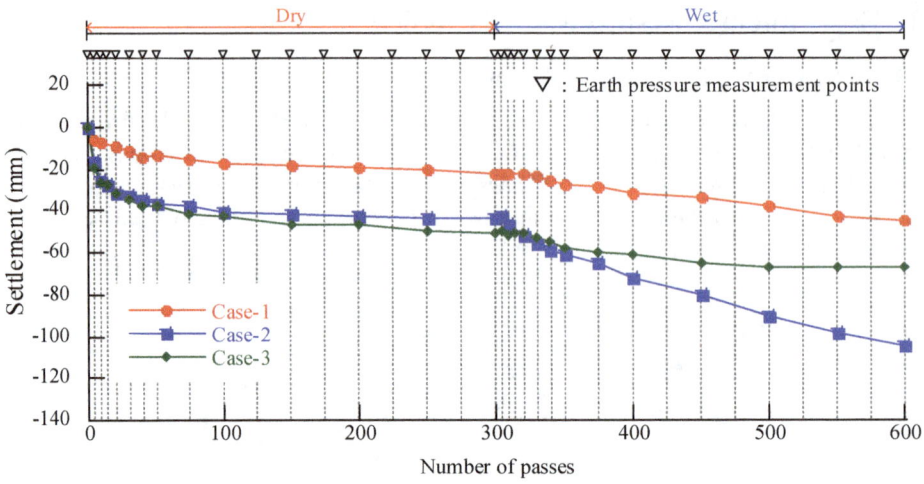

Fig 7. Relationship with settlement and number of passes (0–600)

settlement in Case-2 shown in Fig. 8. In Fig. 8, the settlement after 300 passes for all the cases were set as equal, 0 mm, to compare the settlement caused with the traffic load after sprinkling water. The deformation modulus of base course consist of mechanically stabilized crushed stones is decreased as the water contents increases (Liu et al. 2016). Therefore, the base course in wet condition of Case-1 and 2 were settled. Especially for Case-2, due to the low compaction degree, the larger voids were filled with water and effective stress was decreased, then bigger settlement occurred. On the other hand, in Case-3, due to the reinforcement of Do-nou bags, the settlement was minimized among all the cases.

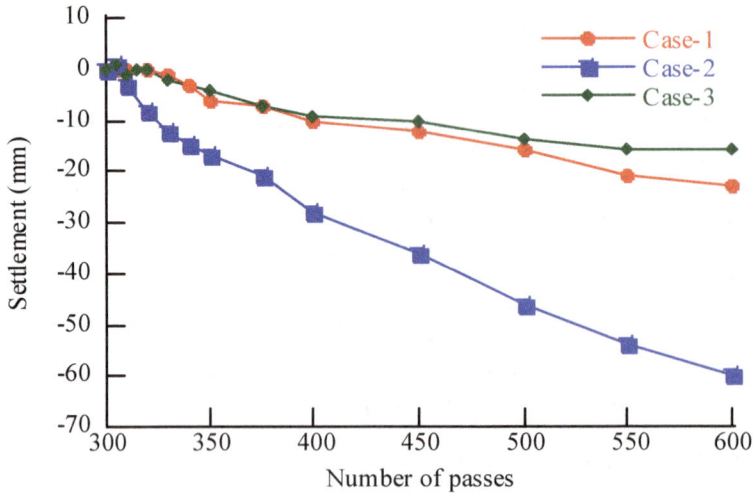

Fig 8. Relationship with settlement and number of passes (300–600)

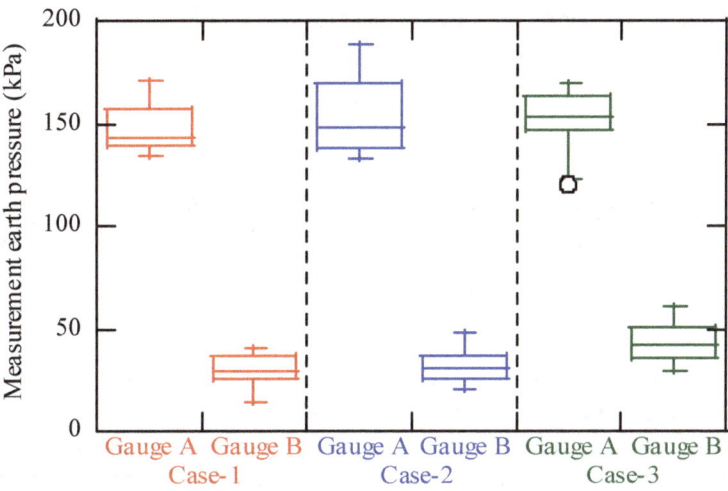

Fig 9. Measurement result of Earth pressure

3.2 Deformation of Base Course

After the final, 600 passes of the 2-ton dump truck, the settlement of the base course surface was measured. Then, in Case-3, the surface layer of the mechanically stabilized crushed stones on the second layer of Do-nou were removed, then the settlement of the surface of the second Do-nou layer were measured. Similarly, in all the cases, all the structure f base course was removed, then the subgrade was exposed to measure the settlement after the driving tests. Table 4 shows the maximum settlement of the base course surface, the settlement of subgrade just below the measuring point of base course surface. With these data, the base course deformation was calculated as shown in Table 4. In Case-1 where the compaction degree reached 97% by compacted using a pedestrian roller, the settlement and deformation were the smallest. When the compaction equipment is not available, Do-nou method application could make the settlement and deformation of base course close to those of the base course constructed with equipment. After the driving tests, DCP tests were conducted at the tire track, then it was found that the bearing capacity of the base course became almost same among all the cases due to the compaction applied through wheel loads during 600 passes (Fig. 4(b)). Though base course settlement and deformation were varied based on the cases as shown in Table 4, it was not found the significant differences from the results of the DCP tests conducted at the tire tracks after the driving tests.

3.3 Load Distributions Analyzed Based on the Measured Earth Pressure

The measurement of the earth pressure gauges placed on the subgrade as shown in Fig. 5 showed the increase with the approach of the 2-ton truck, and then, the maximum value when the tire stepped just above the gauges in longitudinally. With movement away

Table 4. Settlement of base course surface and deformation of base course

	Dry Case-			Wet Case-			Total Case-		
	1	2	3	1	2	3	1	2	3
Settlement of base course surface (mm)	22	44	51	23	60	16	45	104	67
Settlement of subgrade (mm)							12	18	9
Base course deformation (%)							14.7	39.6	17.7

from the location where the gauges were installed, the measured value approached to 0. The measurement with earth pressure gauges were conducted at the designated number of passes as shown in Fig. 7. All the measured values during the driving tests were analyzed with box-and-whisker diagram as shown in (Fig. 9). The measured values of earth pressure gauges during the driving tests were fluctuated.

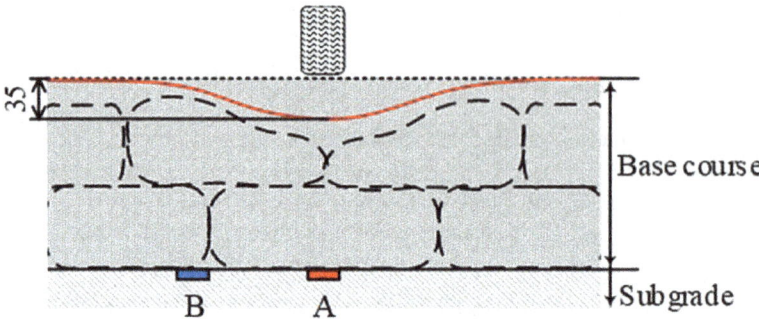

Fig 10. Definition of settlement

In order to compare the load distributions in the base course structure in all the cases, the measured value of the earth pressure gauges when the 35 mm settlement were observed were extracted (Fig. 10) and plotted in Fig. 11. The measured value of Gauge A in Case-3 was the smallest and the ones of Gauge B in Case-3 was the largest. At the same settlement, the earth pressure measured at just below the tire track and the measurement at the location 20 cm away in cross directions were differed according to the structures of base course as shown in Fig. 11. Based on the measured value in Fig. 11, the load distribution of each base course can be analyzed as shown in Fig. 12. When Do-nou method was applied, the load applied to subgrade could be reduced compared to the conventional base course structure.

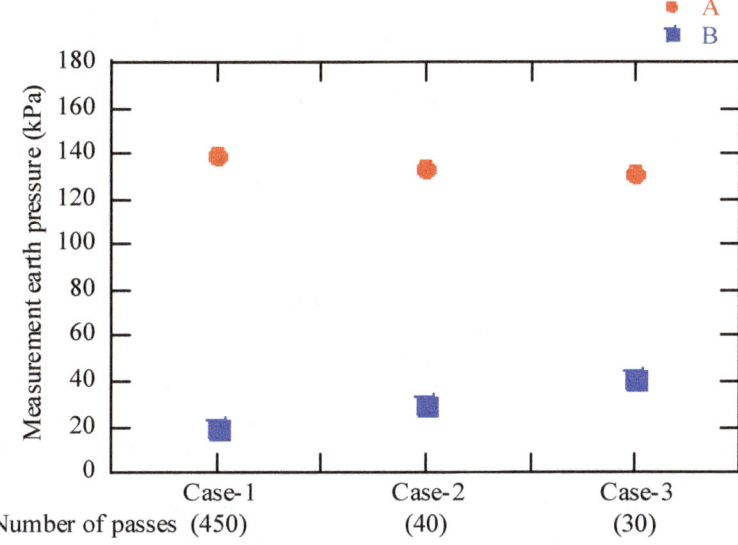

Fig 11. Measurement of earth pressure when the settlement reached to 35 mm

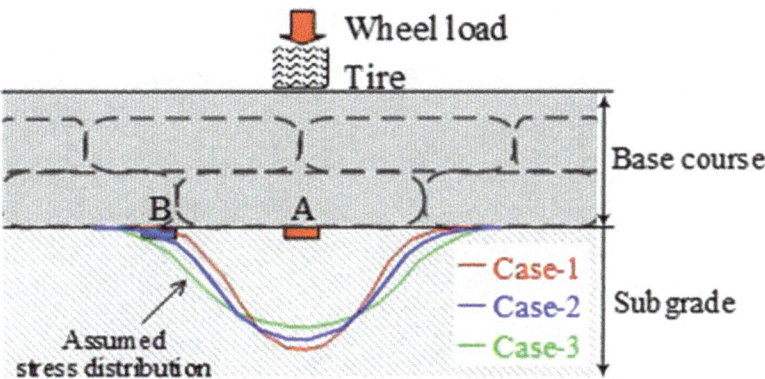

Fig 12. Analyzed stress distribution in each case

4 Conclusions

In this paper, the performance of base course consist of Do-nou was evaluated in comparison to the conventionally designed base course. When the base course was in dry condition, the significant advantages of use of Do-nou method were not found, while in wet condition, it was found that the base course with Do-nou and manual compaction performed similarly to the based course compacted with equipment. Where compaction equipment is not available, Do-nou method with only manual compaction could be alternative of building base couse.

Acknowledgments. The authors would like to gratefully acknowledge the member of the Field Science Center at the University of Miyazaki in Japan for providing the spaces for the driving tests. us a chance to borrow the field for the full-size driving tests. We also wish to express our sincere gratitude to The Geo-science Center Foundation for financial support to this research since 2018 to 2020.

References

Donnges, Ch.; Edmonds, G., Johannessen, B.: Rural Road Maintenance - Sustaining the Benefits of Improved Access-, ILO (2007)

Johannessen, B.: Building Rural Roads, ILO (2008)

Fukubayashi, Y., Kimura, M.: Improvement of rural access roads in developing countries with initiative for self-reliance of communities. Soils Found. **54**(1), 23–25 (2014)

Matsuoka, H., Liu, S.: New earth reinforcement method by soil bags ("DONOU"). Soils Found. **43**(6), 173–188 (2003)

Fukubayashi, Y., Kimura, M.: Spot-improvement of rural roads using a local resource-based approach: case studies from Asia and Africa. Int. Dev. Intech. (2017). https://doi.org/10.5772/66109

Ethiopian Roads Authority: Design manual for low volume roads (2011)

Paige-Green, P., Du Plessis, L.: The use and interpretation of the dynamic cone penetrometer test, CSIR Built Environment, Pretoria (2009)

Bao, L., Qian, S., Duc-Phong, P.: Influence of water content on strength and deformation properties of graded crushed stone under both static and dynamic stress conditions. Electron. J. Geotech. Eng. **21**, 6827–6840 (2016)

Updated Seismic Input for Next Generation of the Egyptian Building Code

Mohamed ElGabry[✉] and Hany M. Hassan

National Research Institute of Astronomy and Geophysics (NRIAG), Helwan, Cairo, Egypt
elgabry@nriag.sci.eg

Abstract. Egypt is prone to earthquakes of moderate and strong size from inland and distant seismogenic sources. The seismicity and seismic hazard studies have indicated that the country face moderate seismic hazard. Current practice in the Egyptian building code considers seismic load using seismic input from old seismic zonation map that go back to the late 90's. Although, this map was developed using a traditional seismic hazard approach, the seismic design strategy, as well as the building code and its update; rely upon it. Also, despite the several developments in the understanding of seismicity, crustal deformation and earthquake mechanics in Egypt that have been achieved in the last two decades, also the presence of theoretical and practical developments, the current seismic load computation of the code is still dangerously adopt and recommend this map. Therefore, the current practice needs an urgent revisit and update. The main aim of this work is to estimate the ground motion parameters in the form of accessible digital maps or databases (e.g. Peak Ground Acceleration, velocity, displacement, response spectra, and time histories) that can be used as input for the next generation of the Egyptian Building code. Physics-based seismic hazard maps and/input are computed where uncertainties are properly evaluated and communicated to the end users. Also, the newly developed maps or input provide a wide range of ground motion parameters in the form of digital databases, not only limited to peak ground motion values, but comprise also excitation records, response spectra or any other ground motion parameter relevant to seismic engineering. The main delivers of this work will be archived and presented using GIS tool that covers whole Egypt with several thousands of sites (nodes), at each node/site. A geospatial database of computed time histories (seismograms and accelerograms) due to all possible seismic sources that can affect those sites are provided. In addition, at each site a representative response spectrum that envelopes all possible ground motion parameters with different percentiles are given.

1 Introduction

For most countries, building codes, either for construction and urban planning, define the minimum requirements or recommendations for a "good performance" in the face of a possible future seismic action that may occur during the life-time of a given construction (e.g. residential buildings, infrastructure, lifeline).

Generally, the concept of Performance Based Design (PBD) consists in designing a system so that it behaves in a desirable way when subject to a certain action (e.g.

H. Shehata and S. El-Badawy (Eds.): *Sustainable Issues in Infrastructure Engineering*, SUCI, pp. 55–79, 2021.
https://doi.org/10.1007/978-3-030-62586-3_5

earthquakes, wind). The key points in such procedure are the identification of a procedure which relates the behavior/response of an object to the applied action, the identification of the limit beyond which the behavior of the object is unacceptable and the correspondent strength of that action. The Performance Based Seismic Design (PBSD) is the application of the PBD in the field of earthquake resistant structure design. Seismic design codes have been developed since the beginning of 1900 in Italy, U.S. and Japan (BSSC 2015; Fasan 2017). The main purpose for which is to mitigate a building's collapse due to earthquake impact, which was evaluated, as introduced in Italy in 1909, through applying lateral forces proportional to the gravitational load of the building. This is the origin of the lateral force method still used today. Such a procedure, neglecting for a moment the problem of the definition of the seismic load, was merely focused on the collapse prevention.

These considerations arise from the fact that, after some small earthquakes, evidence showed that even although buildings did not collapse an extensive non-structural damage was observed.

At present, a modern PBSD process includes mainly the following steps (Bertero and Bertero 2002):

- Seismic Hazard Assessment (SHA);
- Definition of Building Performance Levels (PLs);
- Selection of acceptable Performance Objectives (POs);
- Structural analysis and POs check.

The concept of PBSD was first imbedded into building design guidelines in 1978 with the ATC-06 publication (ATC 1978). The assessment of seismic hazard was based on a single map and the achievement of adequate performance of a building under risk was reached by classifying the buildings in four different Seismic Performance Categories. Each category requires different levels of security and anti-seismic details. The hazard map was defined probabilistically since it was *"policy decision"* of the ATC-06 committee that *"the probability of exceeding the design ground shaking should – as a goal - be roughly the same in all parts of the country"* and *"there is no workable alternative approach to the construction of a seismic design regionalization map which comes close to meeting the goal"* (ATC 1978) even if it was well recognized that the *"assumption* [of Poissonian distribution of earthquakes in time and space] *is of limited validity"*. Hence, for consistency with a priori decisions, the ATC-06 committee adopted a method known to be based on wrong assumptions (Fasan 2017). It worth to say, at that time deterministic seismic hazard approaches were developed and well known *"based upon estimates of the maximum ground shaking experienced during the recorded historical period without consideration of how frequently such motions might occur"* but *"considering the significant cost of designing a structure for extreme ground motions, it is undesirable to require such a design unless there is a significant probability that the extreme motion will occur"*. In other words, ATC committee decided to lower the ground motion level just for cost reduction, so the probabilistic method was used to give an appearance of rationality to the choice does not matter how robust the method was.

This position is supported by the fact that a seismic hazard map was first drafted for ATC 3-06 *"having literally been drawn by a committee"* based on expert judgment and subsequently since this map *"appeared to agree reasonably well with the level of acceleration determined by Algermissen and Perkins [...] their map was used as a guide for the rest of the country"*. It happened that the map of Algermissen and Perkins (Algermissen and Perkins 1976) was based on a "mean return period" of 475 years, so a 10% probability of exceedance in 50 years map was adopted in ATC-06 and subsequently became a standard number all over the world. So the use of an "average life" of 50 years is explained as *"a rather arbitrary convenience"* and the 10% probability of exceedance as a number often taken by statistician *"to be meaningful"* (Bommer and Pinho 2006).

Historically, the same target probability of exceedance *PEY* of 10% in 50 years has been used worldwide as a reference to design ordinary buildings without any validation or clear risk-based rationale and regardless of differences with the U.S.A. in terms of seismicity, construction practices and economic prosperity (Bommer and Pinho 2006). Consequently, the computed ground motion values, as it could be expected given their probabilistic nature, have been systematically exceeded by earthquakes occurred after the publication and adoption of the standard probabilistic hazard map as pointed by Kossobokov and Nekrasova (2012). Moreover, the comparison between different probabilistic hazard maps computed at regional or national scales reveals how the peak ground motion values (e.g. PGA with $PEY = 10\%$ in 50 years) are not consistent from map to map of the same area of interest, and large differences have been found (Nekrasova et al. 2014; Hassan et al. 2017a) due to many factors (e.g. preconception of the seismic hazard analyst, assumptions). Hassan et al. (2017a) indicated that the comparison among different probabilistic seismic hazard maps for Egypt (at national and local scales) for the same site reveals that the PGA values are not consistent and considerable differences are found; the local studies can be more detailed, though not necessarily more reliable (e.g., Klügel 2008). For example, the PGA values for Aswan vary in the range from 0.005–0.2 g, showing a large scatter in the expected values. So, it could be difficult for the potential users to decide what ground motion value to rely on in the design or retrofitting of the built environment. Furthermore, there are substantial differences between PGA values determined by the same study at the same site, but using different GMPEs (e.g., Abdel-Fattah 2005). Those observations and other engineering considerations have resulted in, in some countries (e.g. U.S.A), a change in the value of *PR* from 10% to 2% in 50 years (Fasan 2017), *"In part, 2% in 50 years was selected because USGS had already produced maps for this hazard level"* (BSSC 2015).

The main contribution in the development of the PBSD philosophy of design has been done by the Vision 2000 report (SEAOC 1995) which firstly introduced a multi performance levels check. This report defines a series of performance levels (in terms of acceptable level of damage) that a building should achieve during earthquake actions of different strength. These performance levels are usually defined as: Operational Limit (OL), Immediate Occupancy (IO), Life Safety (LS) and Collapse Prevention (CP). The "mean return periods" arbitrarily associated with them are 43, 72, 475 and 970 years and correspond to a probability of exceedance of 69%, 50%, 10% and 5% in an interval of 50 years, respectively (Fig. 1). The mean return periods corresponding to the four performance levels have been arbitrarily selected for California (Bertero and Bertero 2002) and it has never been motivated (Bommer and Pinho 2006).

Nowadays, the most advanced seismic codes change what is called the reference average life (*Y*) with the change of the importance of the structure (risk category), which

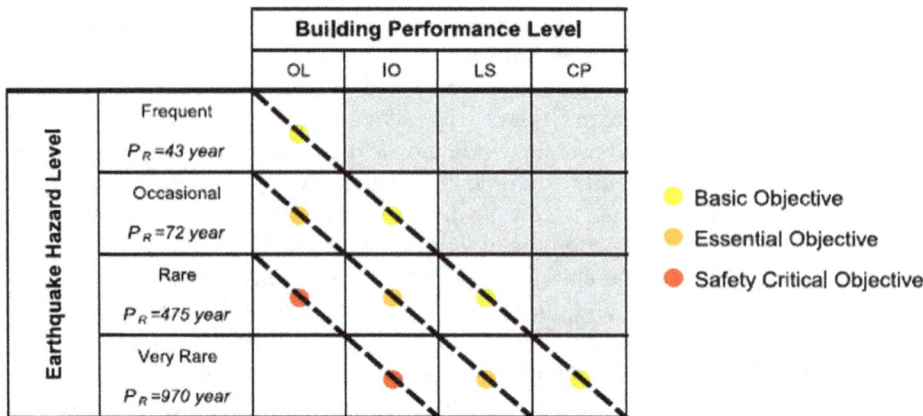

Fig. 1. Vision 2000 conceptual performance objectives matrix (SEAOC 1995).

is controlled by the (hypothetical) consequences of its failure (the more dangerous the consequences, the longer the "average life") (e.g. NTC08 (C.S.L.P. 2008)). This leads the codes to increase the expected structural performance with increasing importance of the structure. Indeed, proper structural performances in case of occurrence of frequent earthquakes (i.e. for low intensities) and, on the other hand, high damages for a very rare earthquake can be accepted.

However, the standard breakdown of the overall cost of a modern building is: 8–18% for structural components, 48–62% for non-structural components and 20–44% for contents (Miranda and Aslani 2003). According to Fasan (2017) the costs optimization using a probabilistic value of ground motions when evaluating the Collapse Prevention Level appears to be unreasonable, at least for three reasons: 1) the fallacy of the "mean return period" concept; 2) the benefits (reduction of costs) due to a PSHA decrease of ground motion involve a very small percentage of the overall cost (the structural components); 3) it does not take into account the post-earthquake recovery costs.

Actually, the PBSD procedure should aim to build an earthquake resilient system. An earthquake resilient system is a system with the following characters (Bruneau et al. 2003):

- reduced failure probabilities of a system;
- reduced consequences from failures, in terms of lives lost, damage, and economic and social consequences;
- reduced to recovery time by increasing resilience (restoration of a specific system or set of systems to their design level of performance).

In fact, the losses due to earthquakes occurred during this decade (e.g. Christchurch (New Zealand) and Tohoku (Japan) earthquakes in 2011), demonstrated that a PBSD procedure based on PSHA is neither reliable nor cost effective. The acceleration response spectra nor 2500 years return period prescribed by the New Zealand seismic code was exceeded by the Christchurch earthquake (22 February 2011, $M_w = 6.2$) that caused 181 deaths. It was estimated that at least 900 buildings in the business district and over 10 thousands homes had to be demolished. The restoration cost was estimated in about US$15--20 billion, the highest cost ever caused by an earthquake in New Zealand (Kaiser et al. 2012; Morgenroth and Armstrong 2012). In the Tohoku earthquake (Japan, 11 March 2011, Mw = 9), followed by a devastating tsunami that cost the government about US$260 billion (Iuchi et al. 2013). The Wenchuan earthquake (China, 12 May 2008, Mw = 7.9) resulted in about US$124 billion of direct losses and at least other US$100 billion of indirect losses due to production interruption (Wu et al. 2012). Fasan (2017) pointed that Italy has spent from 1944 to 2012 almost €181 billion, only in public funding, because of earthquakes Code. Between August and October 2016 a sequence of devastating earthquakes struck the country of much higher spectral accelerations than those with a "mean return period" of 2475 years given by the Italian Building Code (Fasan 2017).

Consequently, it seems obvious that the claim of the ATC-06 committee "*considering the significant cost of designing a structure for extreme ground motions, it is undesirable to require such a design unless there is a significant probability that the extreme motion will occur*" is no longer valid or acceptable in order to create a resilient system. In addition, the progresses in engineering knowledge and the developments of new methods and technologies, such as the use of seismic isolation and/or dissipative systems, make the statement obsolete and even incorrect. According to Sawires et al. (2016a) first version of the Egyptian Concrete Code was developed in 1930, and then it was followed by successive updates in 1962, 1969 and 1988…etc. The Egyptian building codes lacked the guidelines for seismic action till 1988. According to that buildings were typically designed to stand gravity (vertical) and wind (lateral) loads. The first building code of practice to admit and recommend to consider seismic actions in the designing of buildings was published by the Ministry of Housing, Utilities and New Communities in 1988, the Reinforced Concrete Code as pointed by Sawires et al. (2016a) (ESEE 1988). After the occurrence of 1992 Cairo earthquake ($M_w = 5.8$) the Egyptian Code for Loads and Forces was motivated and issued on December 1993 (ECP-201 1993). More than ten years later, a new code (ECP-20 2004) was developed in order to address the shortcomings presented in preceding standards, particularly in the seismic hazard definition. The 2004 code seems to be a copy from the Eurocode 8 version of 2004 with minor changes for making it applicable in Egypt. Most recently, the Egyptian code gradually introduced ductility concepts and detailing procedures through its successive versions (ECP-201 2004, 2008, 2012), although these aspects of the code still need considerable improvement and development at least from the earthquake loading side (Raheem 2013, Sawires et al. 2016a). Reaheem (2013) indicated that the traditional engineering approach aim to employ equivalent static analysis methods, while current design practice is moving toward an increased emphasis on the nonlinear analysis method. In fact, earthquake time histories may be not crucial for the land use and urban planners, but are of a great importance for structural and technical engineers willing to design a new structure and/or evaluate the seismic performance of the existing built environment, and to investigate the non-linear behavior of soil at the site of interest. So, it is crucial to exploit the current methodologies for modeling the generation and propagation of seismic waves, as done with multi-scenario seismic hazard analysis approaches, can to provide a comprehensive database of computed seismograms for Egyptian territories that suffer from the lack of useful strong motion databases.

NDSHA approach is a scenario-based method for seismic hazard analysis, where realistic synthetic seismograms are used to construct earthquake scenarios. In the NDSHA, the knowledge gaps can be appropriately treated, either by incorporating available data from seismology, paleoseismology, morphostructural, geodetic, and related studies or by performing extensive parametric and sensitivity analysis to better and accurately define and addressing the effect endemic lack of knowledge, analyst preconceptions and insufficient data and uncertainties on the resultant hazard maps.

2   Challenges for Seismic Hazard in Egypt

Egypt is well defined as a relatively moderate seismicity country, although it has experienced strong earthquake effects through history from distant (Hellenic arc, Cyprian arc, and the Dead Sea fault system) and nearby (e.g., North Red Sea, Gulf of Aqaba, Gulf of Suez, South-West Cairo (Dahshur Zone), and the continental margin of Egypt) earthquake sources. The reasons behind the strong risk from modest seismic hazard are the high population density, the proximity of some seismic sources to urban cities, profound effect of the path and local site condition, the deterioration of the buildings, absence of maintenance, and the poor design and construction practice. According to the macroseismic data, the 365 Crete, 1303 Rhodes, 1969 Shadwan Island (entrance of the Gulf of Suez), and 1992 Cairo (Dahshur) events are examples of earthquakes that generated the strongest impacts in Egypt. If earthquakes with similar magnitudes happen shortly, a high seismic risk in Egypt is expected due to the increase of exposure and vulnerability, which are the main elements in the risk concept. Therefore, the reliable seismic hazard assessment (SHA) to mitigate the possible losses in the future is a due.

Although many lessons learned through the time, most of the existing seismic hazard studies for Egypt failed to predict the ground motion parameters for earthquakes had occurred after their publication. The failure is evidenced by merely comparing the expected ground motion parameters by different studies with the macroseismic intensity, which is shown and discussed by Hassan et al. (2017a). The failure may be due to the fact that, to identify the location and characteristics of seismotectonic sources for Egypt, only seismological observations (about 120 years) have been considered, while paleoseismological and Morphostructural Zonation (MZ) investigations or similar studies that are suitable to identify seismotectonic sources that may be active over a time scale that is larger (long recurrence) than the instrumental database time span have been not investigated yet or ignored or unappreciated.

Although the importance of the new developments in SHA methodology and practice, it is worth to mention that, most of the available SHA maps for Egypt are based on the traditional approaches and have not implemented the newly proposed improvements in their computations, so far. Approximately, 80% of all SHA studies conducted until now about Egypt at different geographic scales are based on the traditional PSHA and it is still in use in the construction of newly developed SHA maps at different scales (e.g., EzzElarab et al. 2016; Sawires et al. 2016b) upon which the current Egyptian building code is dangerously based.

It worth mention that, the recently released studies have adopted the traditional PSHA method and mainly focus of the collection, update, and revision of the earthquake catalog rather than to the critical review and improvement of the methodology and other elements that are crucial to reach a reliable, as much as possible, estimate of hazard. Moreover, a recently released PSHA study for Egypt has been done by Gaber et al. (2018) has admitted that *"the update of the PSHA maps due to the occurrence of few earthquakes of moderate size and without any real advancement in methodology or inclusion or development of a new investigation will not cause any significant changes in the ground motion values and the pattern of the isocontour maps"*.

It worth to mention that, most of the available SHA studies for Egypt (e.g., Ibrahim and Hattori 1982; Abdel-Fattah 2005; Mohamed et al. 2012; EzzElarab et al. 2016) have limited the output of the seismic hazard assessment to one or two value(s), i.e., peak ground acceleration (PGA) for the horizontal component and response spectrum (RS) rather than the complete frequency content, effective acceleration, bracketed duration, incremental velocity and damaging potential (e.g., Decanini and Mollaioli 1998; Bertero and Uang 1992). Also, they did not give the due attention, in a sound and physically correct way, to the so-called "site-effects", that may be not persistent when earthquake source changes (Molchan et al. 2011). Actually, the sediments of the Nile Valley and its Delta can have a substantial impact on the polarization (also defined amplification/de-amplification) of seismic waves in the horizontal plane and on ground failure or soil liquefaction (e.g., El-Sayed et al. 2001).

In fact, earthquake time histories may be not crucial for the land use and urban planners, but are of a great importance for structural and technical engineers willing to design a new structure and/or evaluate the seismic performance of the existing built environment, and to investigate the non-linear behavior of soil at the site of interest. So, it is crucial to exploit the current methodologies for modeling the generation and propagation of seismic waves, as done with scenario-based approaches, can to provide a comprehensive database of computed seismograms for Egyptian territories that suffer from the lack of useful strong motion databases.

It seems that one of the major problems in the seismic hazard studies carried out for Egypt, is that how much the used earthquake catalogs are representative of the real seismicity of the study area (e.g., Badawy 1998; Saleh 2005). So, it is required to use all available information (e.g., geodesy) and to plan new comprehensive investigations where crucially necessary (e.g., paleoseismology) to better identify and characterize the seismic sources for Egypt. The appropriate incorporation of aforementioned information is an essential factor in SHA by whatever approach and may help in improving the performance of the SHA maps, since the use of historical earthquake records alone may not yield a hazard map of appropriate performance. Badawy (1998) has mentioned that before the 1960s the earthquake location accuracy is not adequate for the analysis. The catalog used in this study is too short and insufficient to reliably estimate the seismic hazard, mainly when the assessment is carried out using PSHA methods, which strongly depend on the amount of data available (35 years of seismological observations are useless in the hazard estimation because of the undue extrapolation to large earthquake occurrence rate). Also, some of the existing studies do not communicate the characteristics of the earthquake catalog being used (e.g., Sabry et al. 2001).

The second important factor is the ground motion prediction equation (GMPE) in the case of PSHA or DSHA and the lithosphere structure in the case of scenario based approaches. In fact, most of the GMPEs used in the calculation of ground motion parameters for Egypt have been developed for other regions that differ, for instance, in the tectonic setting and crustal structure, thus they can be called as imported GMPEs. The reason behind the adoption of imported GMPEs is insufficient strong motion database, which is not sufficient to construct an empirical relationship for Egypt or to explore and evaluate the suitable GMPE (Hassan et al. 2017a, b).

Moreover, the critical review of the seismic hazard maps for the country by Hassan et al. (2017a) has revealed that there is no a clear and proper communication for the characteristics of the GMPEs being used and sometimes is difficult to figure out how those studies have defined a rock site regarding shear velocity. Abdel-Fattah (2005) for example, gives the priority to the results based upon the imported GMPE of Atkinson and Boore (1995, 1997) because, from his point of view, there is a proper consistency between the local and regional seismicity and tectonics of the region (i.e., North America), for which the Atkinson and Boore (1995, 1997) GMPE was estimated, and the local and regional seismicity and tectonics of Egypt. Moreover, the inclusion of various GMPEs for hazard estimation for Egypt has been done in some studies without particular caution for the possible incompatibility between different equations.

Most of the estimated ground motion maps for Egypt are not validated against the available observation or the macroseismic data. Also, the uncertainties associated with the computation of ground motion parameters are neither sufficiently assessed nor presented to the different potential users. The PGA values estimated using the NDSHA approach represent the upper boundary for the different seismic hazard maps in the different regions in Egypt; thus, they turn out to be conservative and physically reliable. Also, there is no significant change in PGA values from El-Sayed et al. (2001), Mourabit et al. (2014), and Hassan et al (2017b) this may be due to the fact that, NDSHA needs only earthquake catalogs with $M \geq 5$ (during the period from 2001 to 2014 occurred just a few events of moderate magnitude). It is worth to mention that, the computation of NDSHA maps available for Egypt is carried out using the earthquake catalog and no information about the control faults or MZ has been used, so far. Although, the fact that the earthquake catalogs for Egypt used in Mourabit et al. (2014), which is an NDSHA based study, and Mohamed et al. (2012) "PSHA studies" are almost the same, the predicted ground motion values obtained from NDSHA are more significant and comparable with the observed intensities.

Base on the reasons mentioned above, it is essential to resort to a more reliable solution for modeling the generation and propagation of seismic waves (e.g., the structural models and related computation of realistic broadband signals as done with scenario based approaches (e.g. Neo-Deterministic Seismic Hazard Analysis approach NDSHA). In fact, the regional structural models are an important input in SHA computation based on NDSHA and have a profound effect on the resultant ground motion maps, although, all of the existing models for Egypt are too simple, and the revision of the crustal models is needed taking into account all the crustal studies available for different regions of the Egyptian territory, and eventually, to plan new comprehensive studies over a regular grid where crucially necessary.

In spite of, the poor performances and fundamental shortcomings of existing PSHA studies available for Egypt, the seismic design strategy as well as the building code and its update still rely upon the maps from those studies. In order to overcome the limits of design procedures based upon PSHA seismic input (Fasan et al. 2015; Rugarli 2014), it is necessary to resort to a new seismic design strategy based upon the NDSHA definition of the seismic input in Egypt.

3 Methodology

Multi-Scenario Seismic Hazard Assessment approach is best suited to compute the ground motion parameters at 1 and 10 Hz cut-off frequencies for a set of 1D structural models and multi-earthquake scenarios at different spatial scales.

In the framework of default procedure (Fig. 2), the study region is covered by a regular grid (usually $0.2° \times 0.2°$). The earthquake sources are centered in the grid cells that fall within the adopted seismogenic zones, while the computation sites are placed at the nodes of a grid that is staggered by $0.1°$ with respect to the sources' grid. For a site-specific analysis, the computation can be preformed for selected observation points instead of the default grid as shown for the study site. A smoothing procedure for the definition of earthquakes location and magnitude, M, is then applied to account for spatial uncertainty and for source extension. After smoothing only the cells (earthquake sources) located within the seismogenic zones are retained. The smoothing process makes Multi-Scenario Seismic Hazard Assessment approach robust and prevents it from the possible uncertainties in the earthquake catalog, which is not required to be complete for $M < 5$. A double-couple point source is placed at the centre of each cell, with a representative focal mechanism which is consistent with the present-day dominant tectonic regime of the corresponding seismogenic zone. Source depth is taken into consideration as a function of magnitude.

The standard method used in Multi-Scenario Seismic Hazard Assessment is the modal summation technique (MS) that is very fast and provide an accurate simulation of ground motion in far source condition. It can be applied only when epicentral distance is greater than focal depth. So, MS is not appropriate for hazard scenario in near source and near field condition. To overcome this limit the discrete wave number technique (DWN) is used where it is numerically accurate over wide range of frequencies and distances but it is more time consuming than MS. In the near field, the discrete wavenumber (DWN) technique is used (Pavlov 2009), while for epicentral distances larger than the hypocentral depth the modal summation technique is adopted. The DWN in the implementation of Pavlov (2009) gives the full wave field, including all body waves and near field effects.

The structural model of interest is composed by a sequence of homogeneous layers. This allows the separation of the equation of motion into two independent problems that can be solved exactly: the propagation of SH (Love) waves, which have particle motion in the y-direction, and the propagation of P-SV (Rayleigh) waves, with particle motion in the xz plane. The waves are decomposed into either those propagating upward and downward in some layers, or into horizontally propagating waves, which either decay or grow exponentially with depth in the other layers. In the half-space that terminates the structural model at depth, horizontal propagation describes the wave motion and

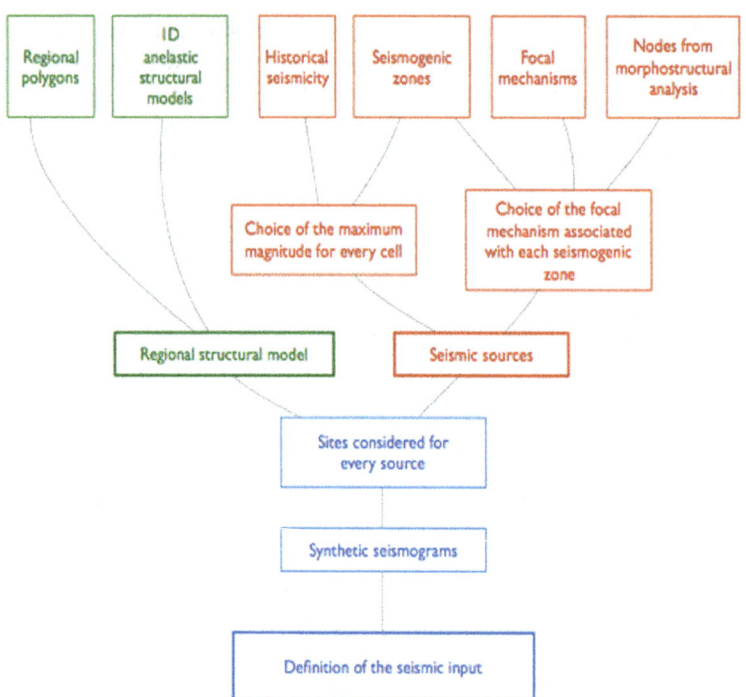

Fig. 2. Flow chart of the different steps in the Multi-Scenario Seismic Hazard Assessment approach for the regional scale analysis

the coefficient of the exponentially increasing wave must vanish. The problem is then reduced to fulfilling all boundary conditions at the interfaces which separate the layers. This leads to an eigenvalue problem in which the eigenvalues (phase velocities) and eigenfunctions (displacement-depth and stress-depth functions) are to be determined.

The seismic source is introduced by using the Ben-Menhaem and Harkrider (1964) formalism. In these expressions, the first-term approximation to cylindrical Hankel functions is used which gives the displacements in the far field. Calculation of synthetic seismograms is then accurate to at least three significant figures, as long as the distance to the source is greater than the wavelength. The seismograms computed in this way contain all the body waves whose phase velocities are smaller than the S-wave velocity of the half-space that terminates the structural model. Starting from the available knowledge about Earth's structure through which seismic waves propagate, seismic sources and seismicity of the study area, it is possible to realistically compute the synthetic seismograms from which one can quantify peak values of acceleration (PGA), velocity (PGV) and displacement (PGD) or any other ground motion parameter relevant to seismic engineering, e.g. response spectra.

4 Input Data for Hazard Computation

In order to perform seismic hazard assessment for the country and for the synthetic seismogram computation, one has to properly model the characteristics of the source generating the seismic energy, and of the structural model through which the seismic wave field propagates.

Earthquake Catalogue
Egypt is well defined as a country of moderate seismicity in a relative sense, although it has experienced strong earthquake impacts along history from regional (e.g. Hellenic arc; Dead Sea fault system) and local (e.g. Gulf of Suez; Dahshur) active earthquake sources even of moderate size (Fig. 3). The high earthquake impacts that happened due the occurrence of moderate size earthquakes can be attributed to a combination of several factors i.e. the proximity of seismic sources to urbanized areas; to the effects of unconsolidated water-saturated soils on the propagating seismic waves over which the cities were built (i.e. horizontal polarization and liquefaction potentialities), the inappropriate engineering design and construction practice of buildings that make the buildings too vulnerable to seismic loads of even modest level. Therefore, the proper inspection of the seismic action and the geotechnical properties of the sediments on which a project will be established are indispensible.

Egypt has a relatively long historical seismic record which goes back about four millennia with variable levels of reliability and detail. The historical seismicity of Egypt has been studied by several authors (e.g., Maamoun et al. 1984; Ambraseys 2009) and the last revised version of the historical catalog was introduced by NRIAG in (2010).

Upgrading a uniform and as much as can completed seismic catalog is an essential step in the seismic hazard analysis. Here, an updated historical and instrumental catalog for earthquakes in and around Egypt (far and near-source) with $M \geq 5$ compiled from different sources covering the period from 2000 BC till 2020 is used. Usually, strong motion from earthquakes with $M \geq 5$ is of much engineering interest due to the possible earthquake impacts (Panza and Suhadolc 1987).

The earthquake catalog that will be used for the computation of the ground motion requires the availability, as complete as possible, only for earthquakes with M5+ . Earthquake catalog for Egypt is updated taking into account the available historical or instrumental earthquake catalogs from national (e.g. Abu El-Nader 2010; Abou Elenean 2009; National Research Institute of Astronomy and Geophysics (NRIAG)) and international sources (e.g. European Mediterranean Seismological Center (EMSC) http://www.emsc-csem.org/; International Seismological Center (ISC) bulletins http://www.isc.ac.uk/isc bulletin/search/). All these information were used to compile a uniform and, as much as possible, complete earthquake catalog. The pre-instrumental earthquake catalog is taken from the revised and quality controlled catalog of NRIAG (2010).

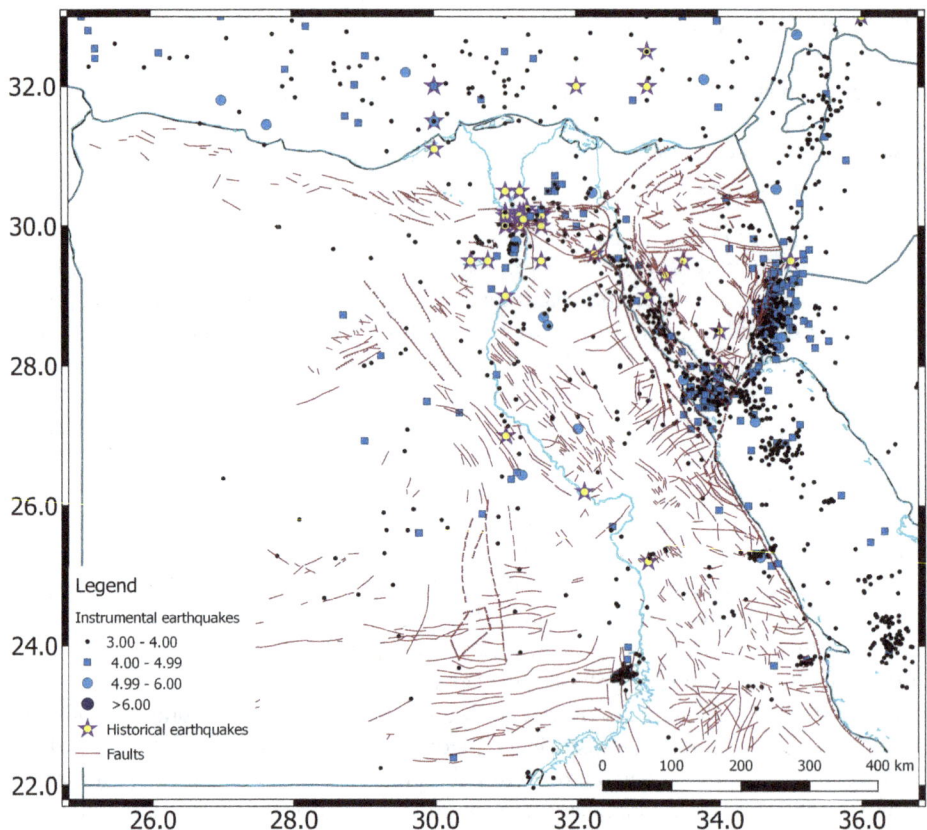

Fig. 3. Historical and instrumental seismicity (from 2000 BC till 2020) and main tectonic features of the study region.

Seismotectonic Model

The principle for seismic source zones is that they represent enclosed areas within which a uniform seismicity distribution and maximum magnitude are expected. Background sources have been avoided in the sense that all areas have been covered by seismic sources, even very low seismicity areas. The principles along which seismic source zones in the current model have been constructed are based on information from geological structures on different scales, tectonics and seismicity.

The upgrade of the seismotectonic model of Egypt was motivated due to newly released information, and it will be subject to future modification when new data are available. Figure 4 shows the seismogenic zones that have been delineated for Egypt using all the available information from different disciplines to be used in the ground motion computation. Most of the defined zones are bounded along well known and clearly defined active trends and faults, while few of them cover the areas that exhibit sporadic seismic activity. The seismotectonic zones, which are marked by strong earthquake M5+ , are identified in this work. This condition could guarantee reliable identification of seismogenic zones not influenced by non-tectonic events (e.g., quarry blasts) which are common in many areas in Egypt. We use magnitude between 3 and 5 to define the zones borders reliably. Then, for areas that did not show earthquake M5+ , we have used pattern recognition analysis for the identification of seismogenic nodes (earthquake prone sites), which represents a complementary step. A responsible seismic hazard assessment requires the incorporation of both results (i.e., defined seismogenic zones and nodes).

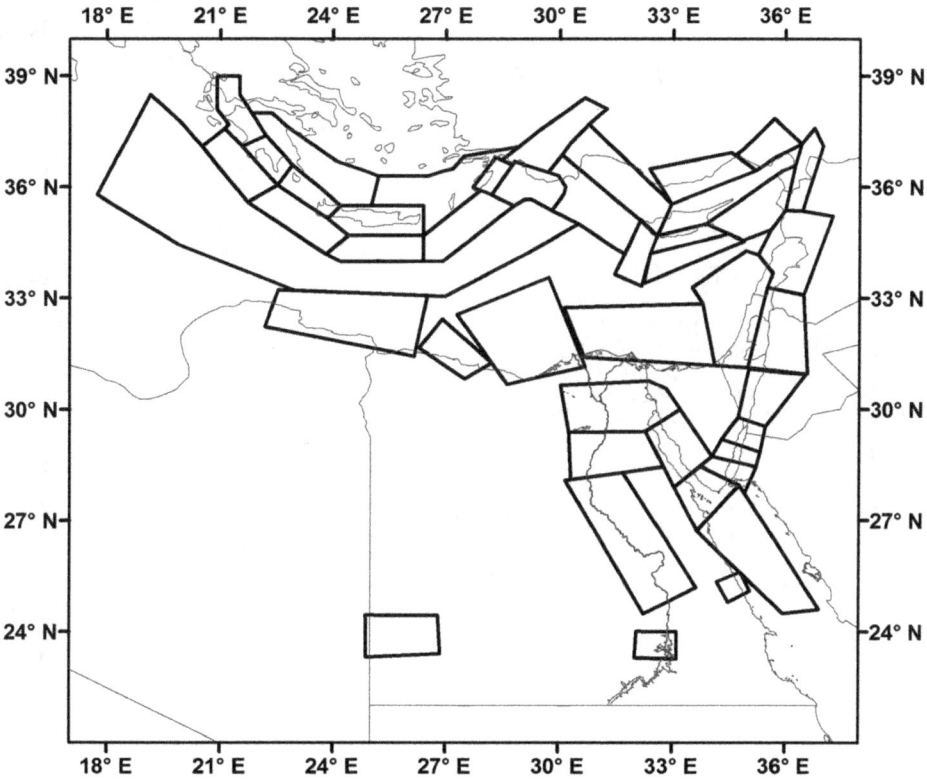

Fig. 4. An updated seismotectonic zones model for the seismic hazard computation at the Ionic Tower site.

A responsible seismic hazard assessment requires the incorporation of both results (i.e., defined seismogenic zones and nodes). Figure 5 displays the delineated nodes developed by Gorshkov et al. (2019), which we are going to be used in hazard computation. We expect that the incorporation of seismogenic nodes information in seismic hazard computation with the seismogenic zones will improve performance of the resulting maps for the studied region, especially for the sites of rare or no seismic activity so far.

Seismic sources are placed in the center of every $0.2° \times 0.2°$ cell that falls within seismogenic zones and seismogenic nodes. The magnitude associated with each source is obtained after a smoothing algorithm (Panza et al. 2001) is applied to the image of seismicity. The smoothing process makes Multi-scenario based seismic hazard approach

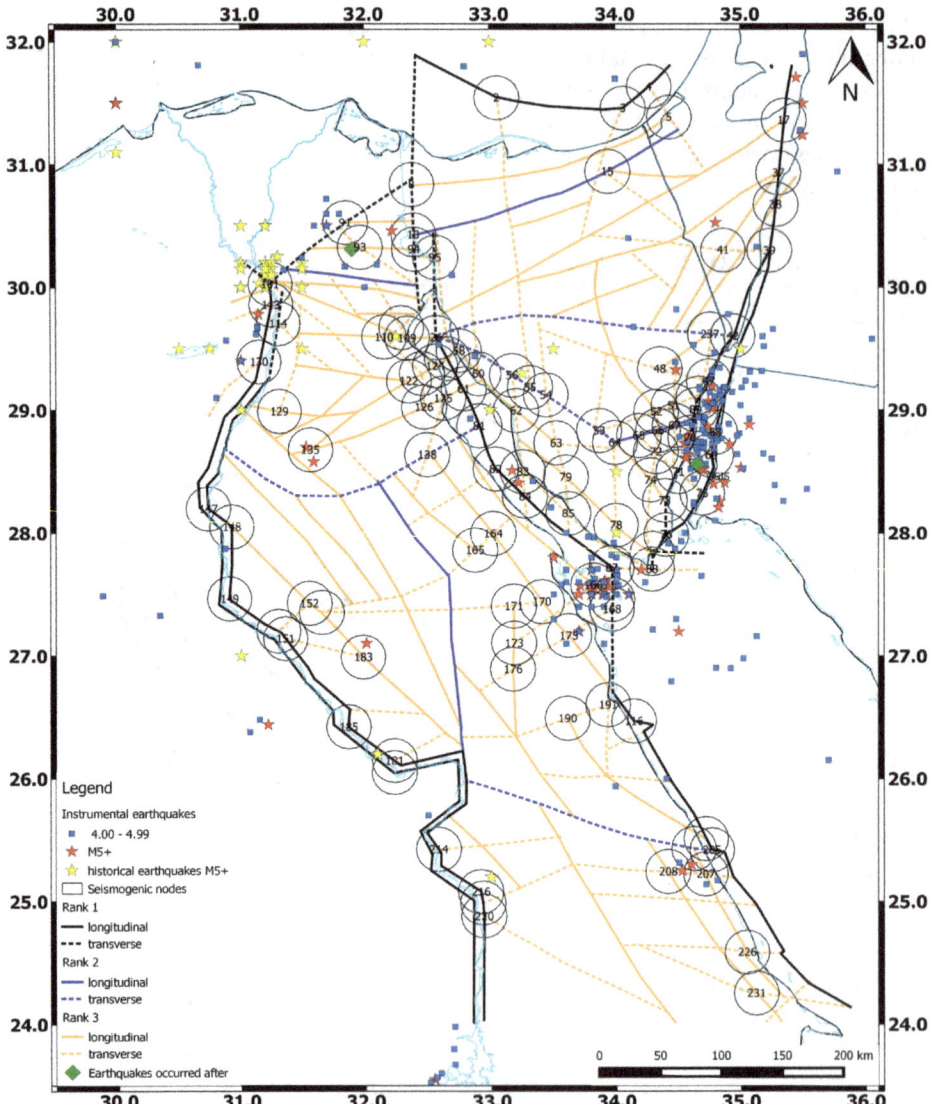

Fig. 5. Seismogenic nodes delineated for easter side of the country by Gorshkov et al. (2019).

robust and prevents the possible uncertainties in the earthquake catalogue, which is not required to be complete for M < 5. This aspect is relevant for the studied area, given that the rather poor completeness of the earthquake catalogue at (M < 5). It also accounts for spatial uncertainties and for source finiteness.

A double-couple point source is placed at the center of each cell, with a representative focal mechanism which is consistent with the present-day dominant tectonic regime of the corresponding seismogenic zone. Source depth is taken into consideration as a function of magnitude (Fig. 6).

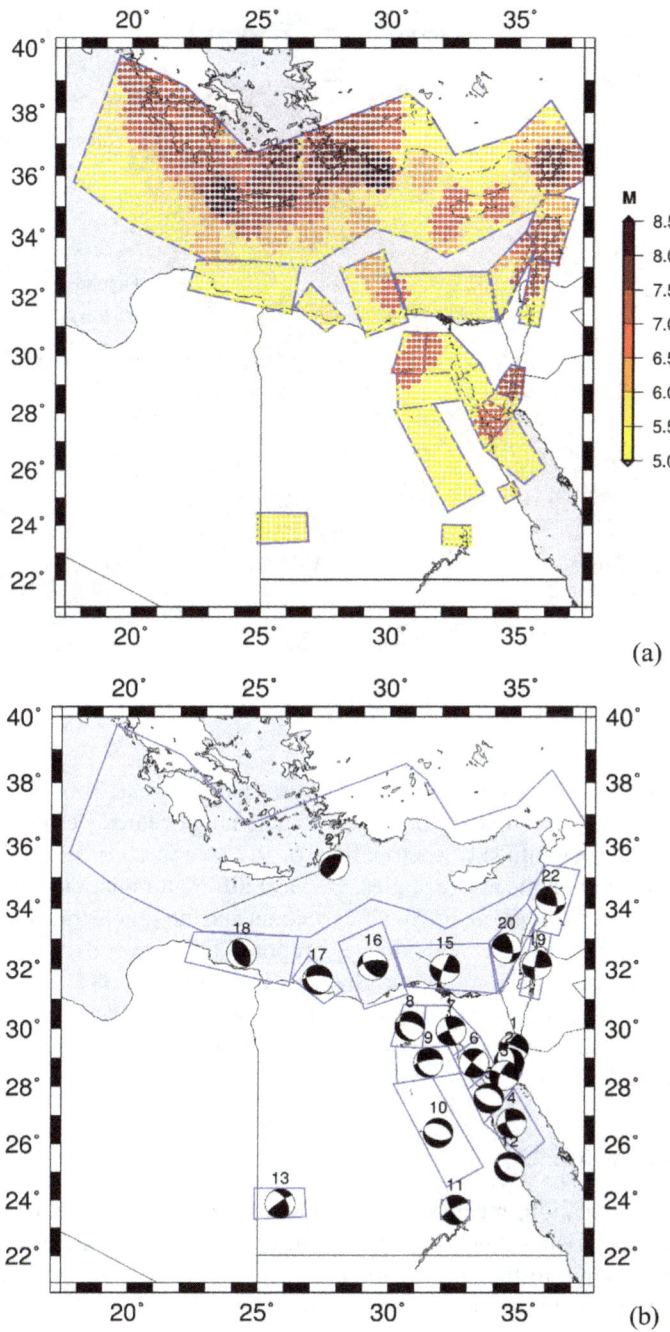

Fig. 6. a) Smoothed magnitude within the seismogenic zones developed in this computation; b) Updated seismotectonic zones and representative focal mechanisms for Egypt; d) Thickness and V_S for the uppermost layer of the updated structural model used in Variant 2.

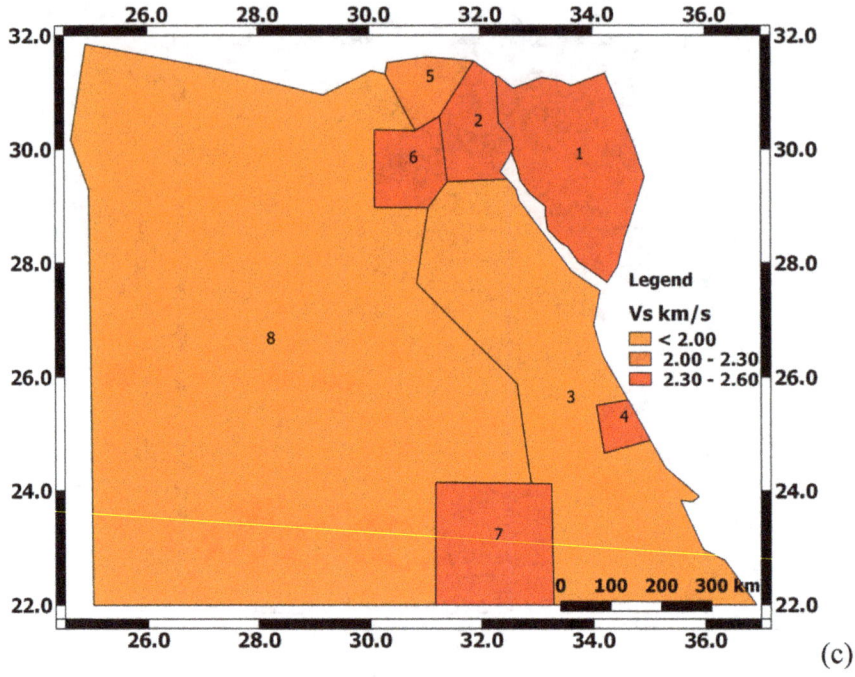

Fig. 6. (*continued*)

Synthetic seismograms are computed in the time and size scaled point-source approximation, using a double-couple of forces placed at the hypocenter. Besides the hypocentral depth and the magnitude, the properties of the source considered are strike, dip and rake. Depending on the above angles, we have different radiation patterns from the source. Directivity effects are also taken into account and the source time functions (STF) are generated according to the methodology proposed by Gusev (2011). Several parameters can be adjusted in generating the Source Time Function STF. We here consider unilateral and bilateral rupturing styles, with backward, neutral and forward directivity with respect to the site position. An example of slip distribution and rupture propagation is shown in Fig. 7.

4.1 Lithospheric Model

Once the source location, mechanism and magnitude have been defined, the Earth's lithosphere characteristics have to be parameterized in order to model wave propagation from each source to the sites of interest. The model is assumed to represent the average properties of the structure between the sources and the sites of interest. The regionalization applied in the area is shown in Fig. 8.

Focusing on the area where the structure is planned, a crustal model has been defined, according to the specific structural properties of the considered region (Fig. 9). The adopted structure model is modified from the 3-D model of seismic P and S velocities in

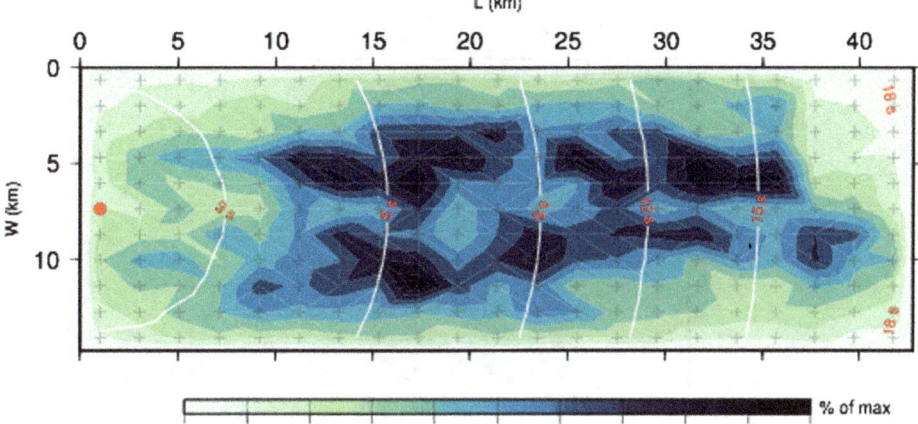

Fig. 7. Slip distribution (green contouring) and rupture velocity (white isochrones) for one realization of the rupturing process of an earthquake with M = 6.9, with unilateral rupture style. Red point on the left side of the figure is the nucleation point

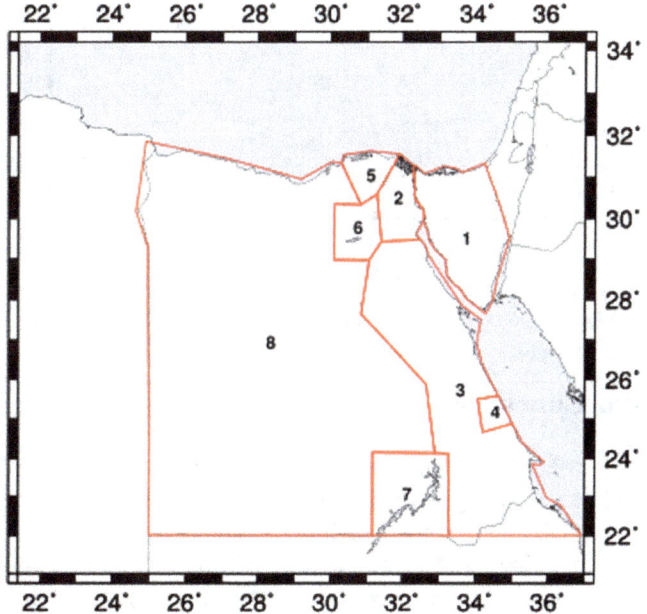

Fig. 8. Regionalization of the crustal properties of Egypt.

the crust and uppermost mantle based on the results of active seismic refraction, which is accomplished by Hassan et al. (2017b).

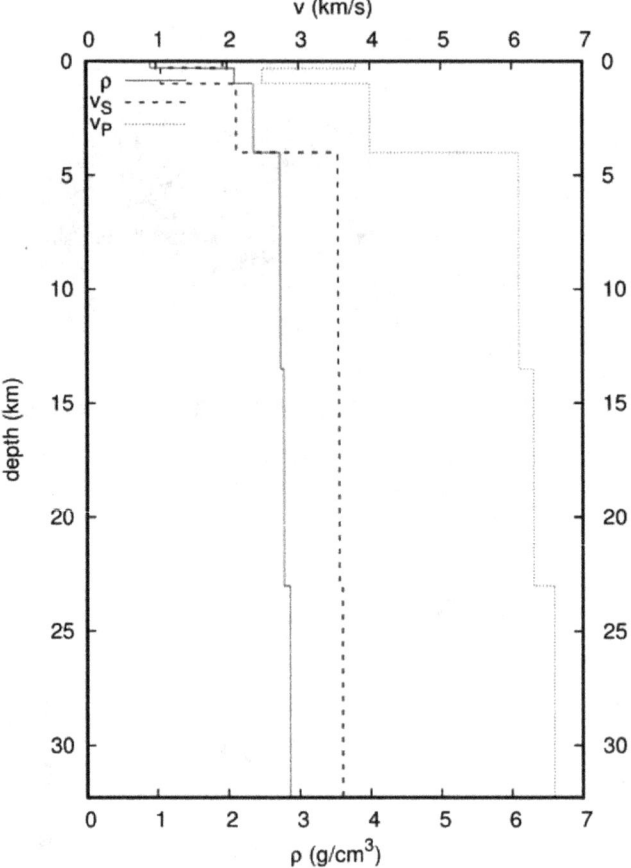

Fig. 9. Layered model associated with polygon n.2 of Fig. 8.

5 Result and Conclusion

In the framework of Multi-Scenario Seismic Hazard Assessment procedure, the study region is covered by a regular grid of sites (0.2° × 0.2°). According to the up mentioned procedures and inputs, ahuge set of sunthitic seismograms is calculated at each grid point all over Egypt bringing at each grid point displacement, velocity and acceleration for every seismic source that may affect that area taking into account all possible uncertainties into the modelling process generating at the end a massive dataset which we processed to generate maps of peak displacement, velocity and acceleration are produced for the whole region of interest (Fig. 10a, b, c).

The ground motion parameters have been computed for the vertical and horizontal components then the maximum value at each site is extracted and plotted as a map of peak ground motion. The PGD, PGV and DGA value at each site are computed and mapped as shown in Fig. 10a, b, c. and present the computed ground motion

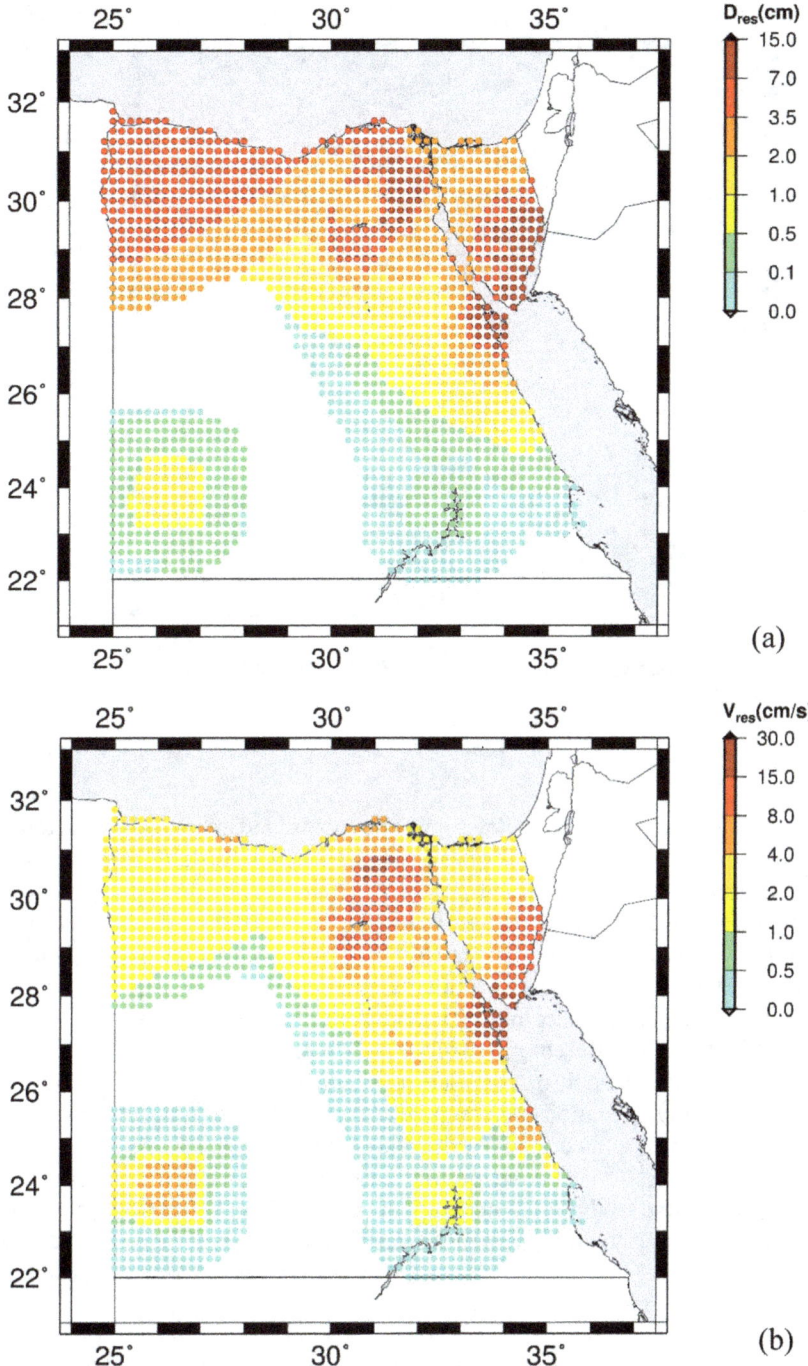

Fig. 10. Peak displacement (a), velocity (b) and acceleration (c) maps for Egypt, for a single realization of the rupturing process at the source

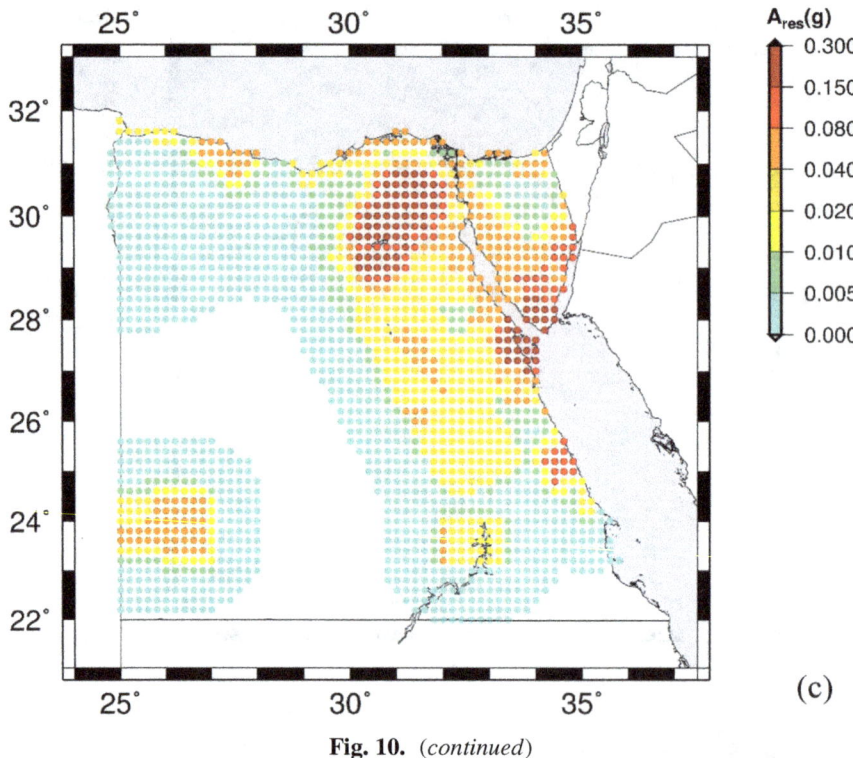

Fig. 10. (*continued*)

For each dot mapped in Fig. 7, the time series computed at bedrock are available and ready to be used as seismic input for the design of seismo-resistant structures or further geotechnical investigations of site response. The map of extracted peak ground motion values shows relatively high earthquake hazard along the Gulf of Aqaba, entrance of the Gulf Suez of Suez, and area around Cairo. These maps can be used directly to pick up the ground motion values instead of smoothed seismic Zonation map.

A designed response spectrum defines the normalized elastic response spectrum of the ground motion for different rock/soil types (i.e., A, B, C and D; they are assigned based on shear-wave velocity), considering a certain percentage of damping can be employed in all the executions to compute the ground motion acceleration at higher frequencies, which is defined for 5% critical damping in future work.

In the updated version of ground motion maps for Egypt, we have revised and updated all input data relative to the reference variant, i.e., earthquake catalog, seismotectonic zones with their representative focal mechanisms and structural models. The release of a new data and data quality control studies, which have been carried out during the last decade has motivated the following updates:

- A single map cannot supply useful enough answers to all the problems posed by adequate seismic risk assessment. The NDSHA approach provides a big database of synthetic seismograms for each configuration, particularly crucial for the regions that suffer from endemic lack of strong motion time histories. The synthetic seismograms is computed with a cut-off frequency as large as 10 Hz for a set of laterally non-varying structural models, where the fault finiteness is duly taken into account by size and time scaled point source (STSPS) model (Parvez et al. 2011), computed with PULSYN06 algorithm (Gusev 2011, Magrin et al. 2016) that provides a broadband kinematic source model.

- Seismic hazard maps should have thorough evaluation of input for hazard computation, methods, and results and the associated uncertainty and validated before they are being accepted then adopted by the code.
- The results obtained represent a database toward detailed and comprehensive ground motion modeling in Egypt.
- An updated building code in order to present and future accommodate theoretical, methodological and application advances in Engineering, Seismology, and Geotechnical aspects of the code is indispensable.

Acknowledgement. This paper and the research behind it would not have been possible without the exceptional support of Prof. Giuliano Panza and his continual support for both authors.

References

Abou Elenean, K.: Earthquake catalog for Egypt. NRIAG internal report (2009)

Abou ELenean, K., Deif, A.: Seismic Zoning of Egypt. Unpublished Work. NRIAG, Egypt (2001)

Abdel Rahaman, M., Tealeb, A., Mohamed, A., Deif, A., Abou Elenean, K., El-Hadidy, M.S.: Seismotectonic zones at Sinai and its surrounding. In: First Arab Conference on Astronomy and Geophysics, October 2008

Abu El-Nader, E.F.: Seismotectonic of Northern Egypt in view of an Updated Earthquake Catalog. Ph.D. thesis, Geology Department, Mansoura, University (2010)

Abdel-Fattah, R.: Seismotectonics of Sinai Peninsula, Egypt and their implications for seismic hazard evaluation. Ph.D. thesis, Faculty of Science, Geology Department, Mansoura University, Egypt (2005)

Algermissen, S., Perkins, D.: A probabilistic estimate of maximum accelerations in rock in the contiguous United States (No. Open-File Report 76–416) (1976)

Ambraseys, N.: Earthquakes in the Mediterranean and Middle East: a Multidisciplinary Study of Seismicity up to 1900. Cambridge University Press, Cambridge (2009)

Ambraseys, N.N.: Value of historical records of earthquakes. Nature **232**, 375–379 (1971)

Ambraseys, N.N.: Far-field effects of Eastern Mediterranean earthquakes in Lower Egypt. J. Seismolog. **5**(2), 263–268 (2001)

Ambraseys, N.N., Bommer, J.J.: The attenuation of ground accelerations in Europe. Earthquake Eng. Struct. Dynam. **20**(12), 1179–1202 (1991)

Ambraseys, N.N., Douglas, J.: Reappraisal of the effect of vertical ground motions on response. Civil and Environmental Engineering Department Imperial College (2000)

Ambraseys, N.N., Melville, C.P., Adams, R.D.: The Seismicity of Egypt, Arabia and the Red Sea: A Historical Review. Cambridge University Press, Cambridge (2005)

ATC: Tentative provisions for the development of seismic regulations for buildings, ATC-3-06 (NBS SP-510). Applied Technology Council (1978)

Atkinson, G.M., Boore, D.M.: Ground-motion relations for eastern North America. Bull. Seismol. Soc. Am. **85**(1), 17–30 (1995)

Atkinson, G.M., Boore, D.M.: Some comparisons between recent ground-motion relations. Seismol. Res. Lett. **68**(1), 24–40 (1997). https://doi.org/10.1785/gssrl.68.1.24

Badawy, A., ElGabry, M., Girgis, M.: Historical seismicity of Egypt. In: A Study for Previous Catalogs Producing Revised Weighted Catalog. The Second Arab Conference for Astronomy and Geophysics, Egypt (2010)

Badawy, A.: Earthquake hazard analysis in northern Egypt. Acta Geodaetica et Geophysica Hungarica **33**(2–4), 341–357 (1998). https://doi.org/10.1007/BF03325544

Ben-Menahem, A., Harkrider, D.G.: Radiation patterns of seismic surface waves from buried dipolar point sources in a flat stratified earth. J. Geophys. Res. **69**(12), 2605–2620 (1964)

Bertero, R.D., Bertero, V.V.: Performance-based seismic engineering: The need for a reliable conceptual comprehensive approach. Earthq. Eng. Struct. Dyn. **31**, 627–652 (2002). https://doi.org/10.1002/eqe.146

BSSC: NEHRP Recommended Seismic Provisions for New Buildings and Other Structures (FEMA P-1050-2). Federal Emergengy Management Agency, Washington, D.C. (2015)

Bertero, V.V., Uang, C.M.: Issues and future directions in the use of an energy approach for seismic resistant design of structures. In: Nonlinear seismic Analysis and Design of Reinforced Concrete Buildings, pp. 3–22 (1992)

Bommer, J.J., Pinho, R.: Adapting earthquake actions in Eurocode 8 for performance-based seismic design. Earthq. Eng. Struct. Dyn. **35**, 39–55 (2006). https://doi.org/10.1002/eqe.530

BSSC: NEHRP Recommended Seismic Provisions for New Buildings and Other Structures (FEMA P-750). Federal Emergengy Management Agency, Washington, D.C. (2009)

Bruneau, M., Chang, S.E., Eguchi, R.T., Lee, G.C., O'Rourke, T.D., Reinhorn, A.M., Shinozuka, M., Tierney, K., Wallace, W.A., von Winterfeldt, D.: A framework to quantitatively assess and enhance the seismic resilience of communities. Earthq. Spectra **19**, 733–752 (2003). https://doi.org/10.1193/1.1623497

C.S.L.P.: Italian Building Code (NTC08). Consiglio Superiore dei Lavori Pubblici (2008)

Decanini, L.D., Mollaioli, F.: Formulation of elastic earthquake input energy spectra. Earthq. Eng. Struct. Dyn. **27**(12), 1503–1522 (1998). https://doi.org/10.1002/(SICI)1096-9845(199812)27: 12%3c1503:AID-EQE797%3e3.0.CO;2-A

ECP-Egyptian Code of Practice-201: Egyptian code for calculating loads and forces. National Research Center for Housing and Building, Ministry of Housing, Utilities and Urban Planning, Cairo (1993)

ECP-Egyptian Code of Practice-201: Egyptian code of practice no. 201 for calculating loads and forces in structural work and masonry. National Research Center for Housing and Building, Ministry of Housing, Utilities and Urban Planning, Cairo (2004)

ECP-Egyptian Code of Practice-201: Egyptian code of practice no. 201 for calculating loads and forces in structural work and masonry. National Research Center for Housing and Building, Ministry of Housing, Utilities and Urban Planning, Cairo (2008)

ECP-Egyptian Code of Practice-201: Egyptian code of practice no. 201 for calculating loads and forces in structural work and masonry. National Research Center for Housing and Building, Ministry of Housing, Utilities and Urban Planning, CairoESEE—Egyptian Society for Earthquake Engineering (1988) Regulations for earthquake resistant design of buildings in Egypt (2012)

Eurocode 8: Design of structures for earthquake resistance. Part 1: general rules, seismic actions and rules for buildings. EN1998-1. European Committee for Standardization, Brussel (2004)

Ezzelarab, M., Shokry, M.M.F., Mohamed, A.M.E., Helal, A.M.A., Mohamed, A.A., El-Hadidy, M.S.: Evaluation of seismic hazard at the northwestern part of Egypt. J. Afr. Earth Sc. **113**, 114–125 (2016). https://doi.org/10.1016/j.jafrearsci.2015.10.017

El-Sayed, A., Vaccari, F., Panza, G.F.: Deterministic seismic hazard in Egypt. Geophys. J. Int. **144**(3), 555–567 (2001). https://doi.org/10.1046/j.1365-246x.2001.01372.x

El-Sayed, A.: Seismic Hazard of Egypt. Ph.D. thesis, Seismological Department Uppsala University, Sweden (1996)

Fasan, M., Amadio, C., Noè, S., Panza, G., Magrin, A., Romanelli, F., Vaccari, F.: A new design strategy based on a deterministic definition of the seismic input to overcome the limits of design procedures based on probabilistic approaches. In: XVI ANIDIS Conference, L'Aquila, Italy (2015). arXiv preprint arXiv:1509.09119

Fasan, M.: Advanced seismological and engineering analysis for structural seismic design. Ph.D. thesis, Trieste University, Italy (2017)

Gaber, H., El-Hadidy, M., Badawy, A.: Up-to-date probabilistic earthquake hazard maps for Egypt. Pure. appl. Geophys. **175**(8), 2693–2720 (2018). https://doi.org/10.1007/s00024-018-1854-5

Gusev, A.A.: Broadband kinematic stochastic simulation of an earthquake source: a refined procedure for application in seismic hazard studies. Pure. Appl. Geophys. **168**(1–2), 155–200 (2011). https://doi.org/10.1007/s00024-010-0156-3

Hassan, H.M., Panza, G.F., Romanelli, F., ElGabry, M.N.: Insight on seismic hazard studies for Egypt. Eng. Geol. **220**, 99–109 (2017a)

Fouad, F.H.: Egypt. In: Paz, M. (ed.) International Handbook of Earthquake Engineering, pp. 195–204. GSHAP-Global Seismic Hazard Assessment Project (1999) (1994). http://www.seismo.ethz.ch/static/GSHAP/

Hassan, H.M., Romanelli, F., Panza, G.F., ElGabry, M.N., Magrin, A.: Update and sensitivity analysis of the neo-deterministic seismic hazard assessment for Egypt. Eng. Geol. **218**, 77–89 (2017b)

Hussein, H.M., Abou Elenean, K.M., Marzouk, I.A., Peresan, A., Korrat, I.M., El-Nader, E.A., Panza, G.F., El-Gabry, M.N.: Integration and magnitude homogenization of the Egyptian earthquake catalog. Nat. Hazards **47**(3), 525–546 (2008). https://doi.org/10.1007/s11069-008-9237-3

Ibrahim, E.M., Hattori, S.: Seismic risk maps for Egypt and vicinity. Helwan Inst. Astronom. Geophy. **2**(Series B), 183–207 (1982)

Iuchi, K., Johnson, L.A., Olshansky, R.B.: Securing tohoku's future: planning for rebuilding in the first year following the Tohoku-Oki earthquake and tsunami. Earthq. Spectra **29**, S479–S499 (2013). https://doi.org/10.1193/1.4000119

Kossobokov, V.G., Nekrasova, A.K.: Global seismic hazard assessment program maps are erroneous. Seismic Instrum. **48**(2), 162–170 (2012). https://doi.org/10.3103/S0747923912020065

Kaiser, A., Holden, C., Beavan, J., Beetham, D., Benites, R., Celentano, A., Collett, D., Cousins, J., Cubrinovski, M., Dellow, G., Denys, P., Fielding, E., Fry, B., Gerstenberger, M., Langridge, R., Massey, C., Motagh, M., Pondard, N., McVerry, G., Ristau, J., Stirling, M., Thomas, J., Uma, S., Zhao, J.: The Mw 6.2 Christchurch earthquake of February 2011: preliminary report. NZ J. Geol. Geophys. **55**, 67–90 (2012). https://doi.org/10.1080/00288306.2011.641182

Kebeasy, R., Maamoun, M., Albert, R.: Earthquake activity and earthquake risk around the Alexandria area in Egypt. Acta Geophys. Pol. **29**(1), 37–48 (1981)

Klügel, J.-U.: Seismic hazard analysis — Quo vadis? Earth Sci. Rev. **88**, 1–32 (2008). https://doi.org/10.1016/j.earscirev.2008.01.003

Maamoun, M.: Observed intensity – epicentral distance relations in Egyptian earthquakes, p. 184. Bull. of Helwan Obs, No (1979)

Maamoun, M., Megahed, A., Allam, A.: Seismicity of Egypt. HIAG Bull. **IV**(B), 109–160 (1984)

Magrin, A., Gusev, A.A., Romanelli, F., Vaccari, F., Panza, G.F.: Broadband NDSHA computations and earthquake ground motion observations for the Italian territory. Int. J. Earthq. Impact Eng. **1**(1–2), 131–158 (2016)

Mohamed, A.E.E.A., El-Hadidy, M., Deif, A., Elenean, K.A.: Seismic hazard studies in Egypt. NRIAG J. Astron. Geophys. **1**(2), 119–140 (2012). https://doi.org/10.1016/j.nrjag.2012.12.008

Molchan, G., Kronrod, T., Panza, G.F.: Hot/cold spots in Italian macroseismic data. Pure. appl. Geophys. **168**(3–4), 739–752 (2011). https://doi.org/10.1007/s00024-010-0111-3

Morgenroth, J., Armstrong, T.: The impact of significant earthquakes on Christchurch, New Zealand's urban forest. Urban For. Urban Greening **11**, 383–389 (2012). https://doi.org/10.1016/j.ufug.2012.06.003

Mourabit, T., Abou Elenean, K.M., Ayadi, A., Benouar, D., Ben Suleman, A., Bezzeghoud, M., Cheddadi, A., Chourak, M., ElGabry, M.N., Harbi, A., Hfaiedh, M., Hussein, H.M., Kacem, J., Ksentini, A., Jabour, N., Magrin, A., Maouche, S., Meghraoui, M., Ousadou, F., Panza, G.F., Peresan, A., Romdhane, N., Vaccari, F., Zuccolo, E.: Neo-deterministic seismic hazard assessment in North Africa. J. Seismolog. **18**(2), 301–318 (2013). https://doi.org/10.1007/s10950-013-9375-2

Musson, R.M.: The effect of magnitude uncertainty on earthquake activity rates. Bull. Seismol. Soc. Am. **102**(6), 2771–2775 (2012). https://doi.org/10.1785/0120110224

Nekrasova, A., Kossobokov, V., Peresan, A., Magrin, A.: The comparison of the NDSHA, PSHA seismic hazard maps and real seismicity for the Italian territory (2014)

Panza, G.F., Suhadolc, P.: Complete strong motion synthetics. Seismic Strong Motion Synth. **4**, 153–204 (1987)

Panza, G.F., Cazzaro, R., Vaccari, F.: Correlation between macroseismic intensities and seismic ground motion parameters (1997)

Panza, G.F., Kossobovok, G.V., Peresan, A., Nekrasova, A.: Why are the standard probabilistic methods of estimating seismic hazard and risks too often wrong. Earthq. Hazard Risk Disasters (2014). https://doi.org/10.1016/B978-0-12-394848-9.00012-2

Panza, G.F., La Mura, C., Peresan, A., Romanelli, F., Vaccari, F.: Chapter three-seismic hazard scenarios as preventive tools for a disaster resilient society. Adv. Geophys. **53**, 93–165 (2012)

Panza, G.F., Peresan, A., Magrin, A., Vaccari, F., Sabadini, R., Crippa, B., Marotta, A.M., Splendore, R., Barzaghi, R., Borghi, A., Cannizzaro, L., Amodio, A., Zoffoli, S.: The SISMA prototype system: integrating Geophysical Modeling and Earth Observation for time-dependent seismic hazard assessment. Nat. Hazards **69**(2), 1179–1198 (2011). https://doi.org/10.1007/s11069-011-9981-7

Panza, G.F., Romanelli, F., Vaccari, F.: Seismic wave propagation in laterally heterogeneous anelastic media: theory and applications to seismic zonation. Adv. Geophys. **43**, 1–95 (2001)

Panza, G.F., Vaccari, F., Cazzaro, R.: Deterministic seismic hazard assessment. In: Vrancea Earthquakes: Tectonics, Hazard and Risk Mitigation, pp. 269–286. Springer, Dordrecht. http://dx.doi.org/10.1007/978-94-011-4748-4_25

Panza, G.F., Romanelli, F., Vaccari, F.: Seismic wave propagation in laterally heterogeneous anelastic media: theory and applications to seismic zonation. In: Advances in Geophysics, vol. 43, pp. 1–95. Academic Press (2001)

Parvez, I.A., Romanelli, F., Panza, G.F.: Long period ground motion at bedrock level in Delhi city from Himalayan earthquake scenarios. Pure. appl. Geophys. **168**(3–4), 409–477 (2011). https://doi.org/10.1007/s00024-010-0162-5

Pavlov, V.M.: Matrix impedance in the problem of the calculation of synthetic seismograms for a layered-homogeneous isotropic elastic medium. Izvestiya Phys. Solid Earth **45**, 850–860 (2009). https://doi.org/10.1134/S1069351309100036

SEAOC: Vision 2000: Performance Based Seismic Engineering of Buildings. Structural Engineers Association of California, Sacramento, California (1995)

Raheem, S.E.A.: Evaluation of Egyptian code provisions for seismic design of moment-resisting-framemulti-story buildings. Int. J. Adv. Struct. Eng. **5**, 1–18 (2013)

Reiter, L.: Earthquake Hazard Analysis. Columbia University Press, New York (1991)

Riad, S., Ghalib, M., El-Difrawy, M.A., Gamal, M.: Probabilistic seismic hazard assessment in Egypt. Ann. Geol. Surv. Egypt **23**, 851–881 (2000)

Rugarli, P.: Validazione Strutturale. EPC (2014)

Sabry, A.A., Agaiby, S.W., Mourad, S.A., Aly, T.M.: Seismic hazard of Egypt with consideration to local geotechnical conditions. Ph.D. thesis, Faculty of Engineering, Cairo University (2001)

Saleh, K.H.: Active faulting and seismic hazard assessment of the north western desert, with contribution of geographic information system. Ann. Geol. Surv. Egypt **28**, 435–560 (2005)

Sawires, R., Peláez, J.A., Fat-Helbary, R.E., Ibrahim, H.A., García-Hernández, M.T.: An updated seismic source model for Egypt. In: Earthquake Engineering—From Engineering Seismology to Optimal Seismic Design of Engineering Structures, pp. 1–51. InTech, Croatia (2015)

Sawires, R., Peláez, J.A., Fat-Helbary, R.E., Ibrahim, H.A.: A review of seismic hazard assessment studies and hazard description in the building codes for Egypt. Acta Geod. Geoph. **51**(2), 151–180 (2016a). https://doi.org/10.1007/s40328-015-0117-5

Sawires, R., Peláez, J.A., Fat-Helbary, R.E., Ibrahim, H.A.: Updated probabilistic seismic hazard values for Egypt. Bull. Seismol. Soc. Am. (2016b). https://doi.org/10.1785/0120150218

Wu, J., Li, N., Hallegatte, S., Shi, P., Hu, A., Liu, X.: Regional indirect economic impact evaluation of the 2008 Wenchuan Earthquake. Environ. Earth Sci. **65**, 161–172 (2012). https://doi.org/10.1007/s12665-011-1078-9

Effect of Using Different Installation Positions for Stirrups in Enhancing the Shear Capacity of RC Beams

Ghada Gamal Salem Ahmed[(✉)]

Construction Engineering Department, Egyptian Russian University ERU, Badr City, Cairo 11829, Egypt
civilghada@hotmail.com, ghada-gamal@eru.edu.eg

Abstract. Reinforced concrete beams are important supporting elements for most of the construction purposes. There are different modes of failure that can happen for beams, but this research will focus on shear failure and how we can resist the diagonal shear crack. Stirrups are an essential reinforcement in beams that can resist the shearing force. It is a closed-loop that holds the main reinforcement bars together, and it is usually installed in a vertical position. The current research presents an analytical study on the effect of using different installation positions for stirrups in beams such as; vertical stirrups (conventional technique), inclined stirrups, and a combination of vertical and inclined stirrups. The analytical study is carried out with the Finite Element Method using a software package (ANSYS APDL). A verification model is carried out at first, and compared with previous experimental work to validate the results, then the following parameters are studied: 1- Stirrups installation positions, 2- Types of Stirrups (single-legged – two-legged) 3- Angle of inclination. All beams are loaded up to failure then the failure loads, deflections, and crack patterns are obtained, also the P-delta curves are drawn for all beams.

1 Introduction

There are two main types of failure that can occur in reinforced concrete beams: flexure failure and shear failure. The shear failure occurs when the beam has shear resistance lower than flexural strength and the shear force exceeds the shear capacity of different materials of the beam. The shear failure causes some diagonal cracks at the shear zone of the beam which should be resisted with the shear reinforcement. Stirrups are the shear reinforcement that have an important role to prevent shear and torsion, beside it is role as a holder for the longitudinal reinforcement in their places as shown in Fig. 1. Stirrups are always installed vertically to be much easier in their installation, but it may be more effective if we can install them in inclined position to be perpendicular with the diagonal shear crack. The inclination of stirrups and their effect in the shear capacity is our aim in the current study.

H. Shehata and S. El-Badawy (Eds.): *Sustainable Issues in Infrastructure Engineering*, SUCI, pp. 80–92, 2021.
https://doi.org/10.1007/978-3-030-62586-3_6

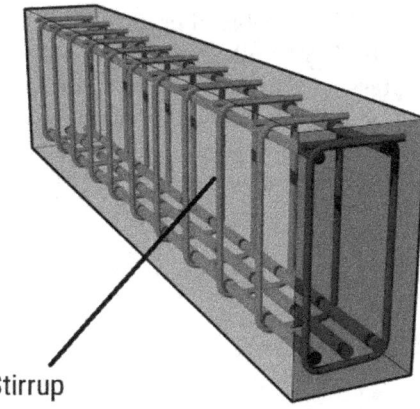

Stirrup

Fig. 1. Traditional shear reinforcement for beams (vertical stirrups)

Many researches presented an experimental study on the effect of use the inclined stirrups to resist shearing force more effectively [1–5] but still in need to study more parameters. Also, some researches have turned to study the effectiveness of using FRP materials as shear reinforcement in beams [6], and others turned to use continuous spiral transversal reinforcement [7].

The previous experimental investigations showed that the inclined stirrups enhanced the ductility and increased the shear capacity of beams.

2 Objective

The main objective of this research is to study the effect of using inclined stirrups in enhancing the shear capacity of RC beams. The study is done analytically using Finite Element Method on software package ANSYS 15.0.

The sequence of this research on ANSYS software is:

1- Verification models by simulating specimens from previous experimental work to validate the ANSYS results at first.
2- Parametric study by simulating new models to study the following parameters:

    a) Installation positions for stirrups (vertical – Inclined – Combination).
    b) Types of stirrups (single-legged – two-legged).
    c) Angle of inclination (30–45–60).

3 Analytical Models and Element Types

The analytical study on ANSYS is done by simulating all specimens with three dimensional elements. Solid65 is used to model the concrete material. The element is defined by eight nodes having three degrees of freedom at each node: translations in the nodal x, y, and z directions. The solid is capable of cracking in tension and crushing in compression. A 3-D element (LINK180) will be used to model reinforcement. The element is a uniaxial tension-compression element with three degrees of freedom at each node: translations in the nodal x, y, and z directions. Tension-only (cable) and compression-only (gap) options are supported. A 3-D solid element (SOLID185) is used to model the steel plates for loading and supports (Fig. 2).

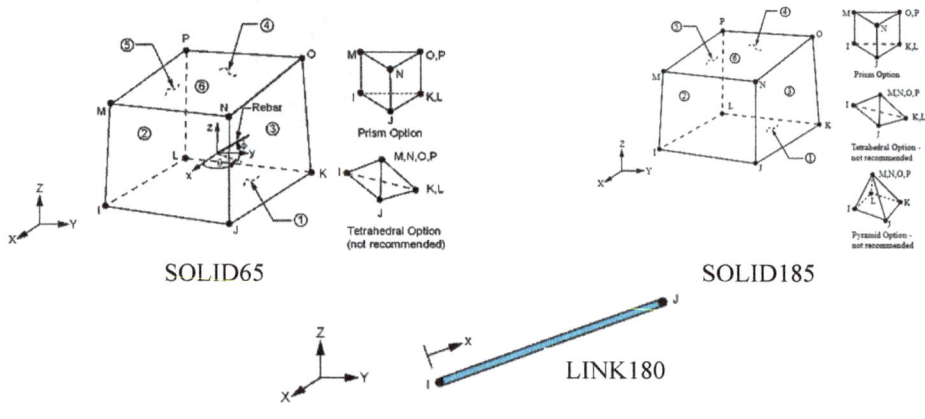

SOLID65   SOLID185

LINK180

Fig. 2. Element types

4 Verification Models

Nor Zamri et al. [1], presented an experimental investigation to study the effect of using inclined stirrups with 45° on increasing the shear capacity of beams. Two of their experimental specimens are used in our study to validate the ANSYS results. These specimens are simulated using the three-dimensional elements mentioned in the previous section. Table 1, Fig. 3 and Fig. 4 illustrates the specimens' details. Compressive strength of concrete after 28 days from Nor's experiments was 46 MPa. High tensile steel with grade of 460 MPa and mild tensile steel with grade of 250 MPa.

Table 1. Specimens' details

Specimen's label	Cross section (mm)	Compression RFT	Tension RFT	Stirrups position	Angle of inclination
CRB-A	150250	2T16	2T10	Vertical	–
CRB-B	150 × 250	2T16	2T10	Inclined	45°

Four concentrated loads were exerted on each beam. Figure 3 illustrates the loads position from each support and it also illustrates the spacing between stirrups along the beam. Figure 4 illustrates the inclination of stirrups for beam CRB-B.

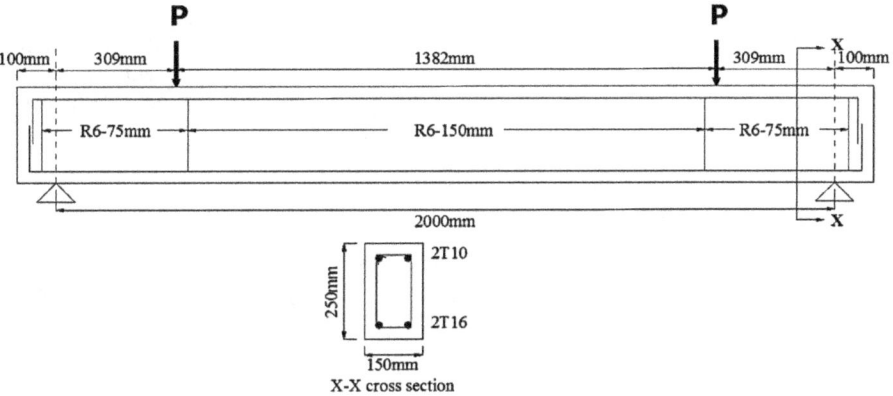

Fig. 3. Specimens' details

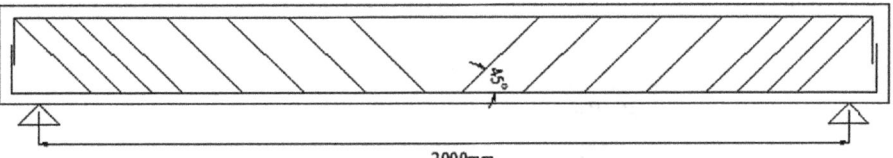

Fig. 4. Inclination of stirrups for CRB-B

5   Parametric Study

A parametric study is carried out on eight beams with different stirrup's installation position as shown in Table 2. Compressive strength of concrete is 30 MPa. High tensile steel with grade of 460 MPa and mild tensile steel with grade of 250 MPa.

Table 2. Models of the parametric study

Group	Beam no.	Stirrup's position	Type of stirrup	Angle of inclination	Notes
1	B1	Vertical only	Two-legged	–	
	B2	Inclined only	Two-legged	45	B2 is in common with Group (1) and Group (3)
	B3	Combination	Two-legged	45	
2	B4	Vertical only	Single-legged	–	
	B5	Inclined only	Single-legged	45	
	B6	Combination	Single-legged	45	
3	B7	Inclined only	Two-legged	30	B2 is also in Group (3) with B7 and B8
	B8	Inclined only	Two-legged	60	

The concrete dimensions and the distribution of stirrups along the beam length is the same as Fig. 3. The stirrup's types are illustrated in Fig. 5.

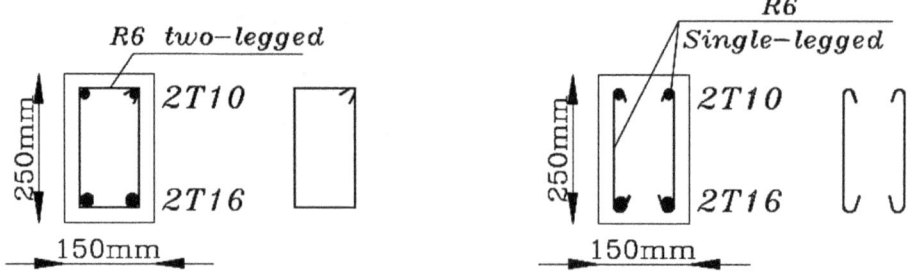

Fig. 5. Types of stirrups (single-legged and two-legged)

The following figures illustrate the simulation of beams with Finite Element Method on ANSYS software. All beams are loaded up to failure then the failure loads, deflections, and crack patterns are obtained, also the P-delta curves are drawn for all beams (Fig. 6).

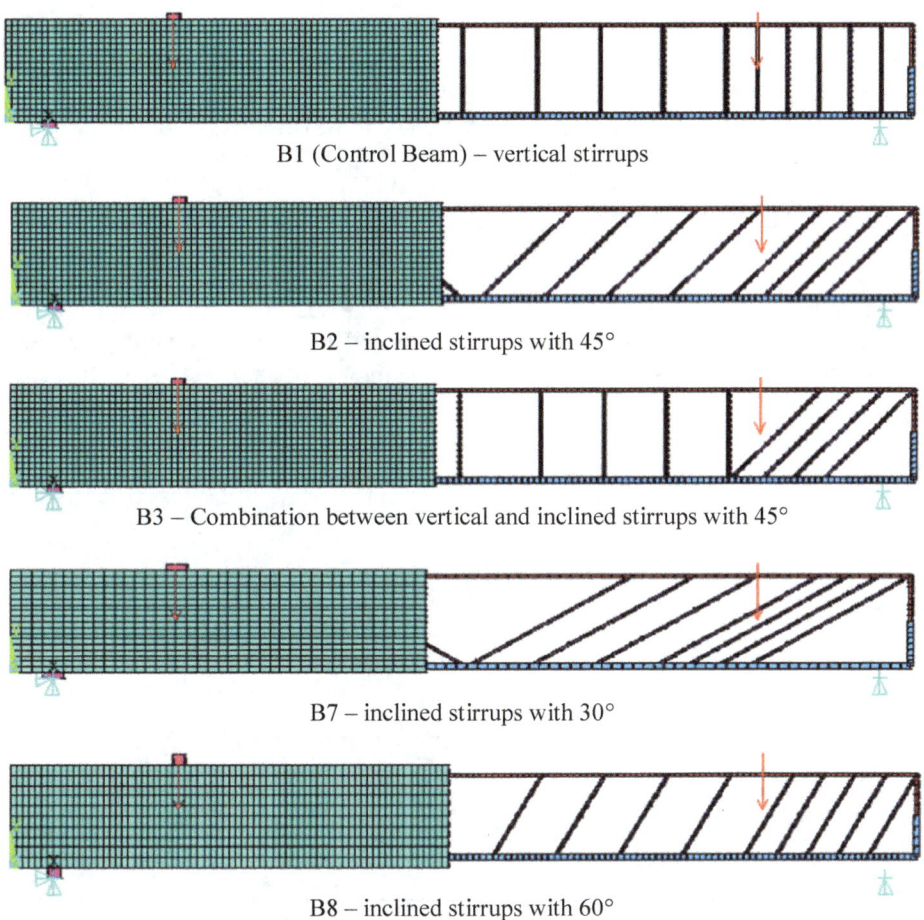

B1 (Control Beam) – vertical stirrups

B2 – inclined stirrups with 45°

B3 – Combination between vertical and inclined stirrups with 45°

B7 – inclined stirrups with 30°

B8 – inclined stirrups with 60°

Fig. 6. ANSYS models

6 Results and Discussion

6.1 Results of the Verification Models

The results from the verification models are recorded and compared with Nor's experimental tests [1] in order to validate our Finite Element Analysis. Table 3 illustrates the failure loads and their corresponding displacement.

Table 3. Experimental results and results of the verification models from ANSYS

Specimen's label	Experimental results		ANSYS results	
	P_{exp} (KN)	Δ_{exp} (mm)	P_{ANS} (KN)	Δ_{ANS} (mm)
CRB-A	207.1	16.8	207	17.5
CRB-B	252.4	11.4	252	8.22

Each beam is loaded gradually up to failure, then loads from each step are recorded with their corresponding displacement to draw the P-Delta Curve, and compare it with the experimental curve as shown in the following figures (Figs. 7 and 8).

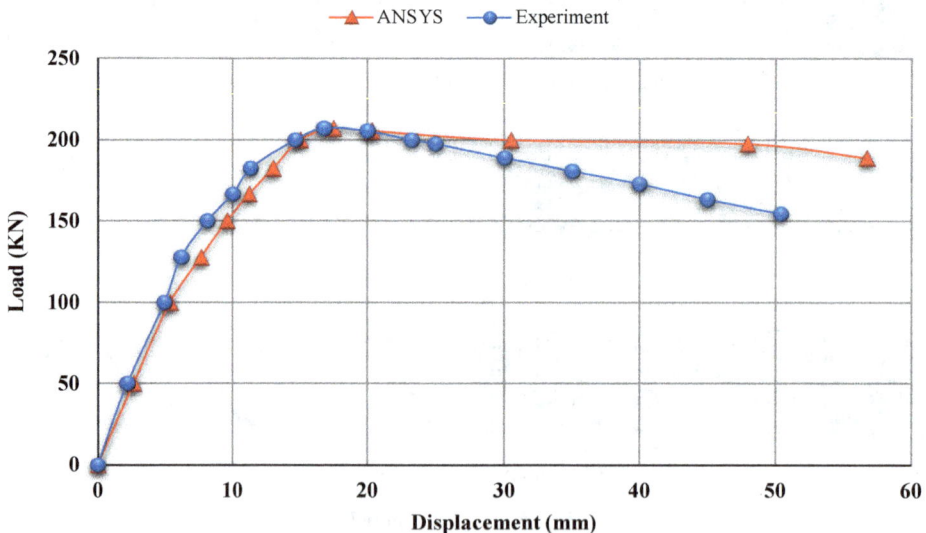

Fig. 7. P-Delta curve for specimen CRB-A

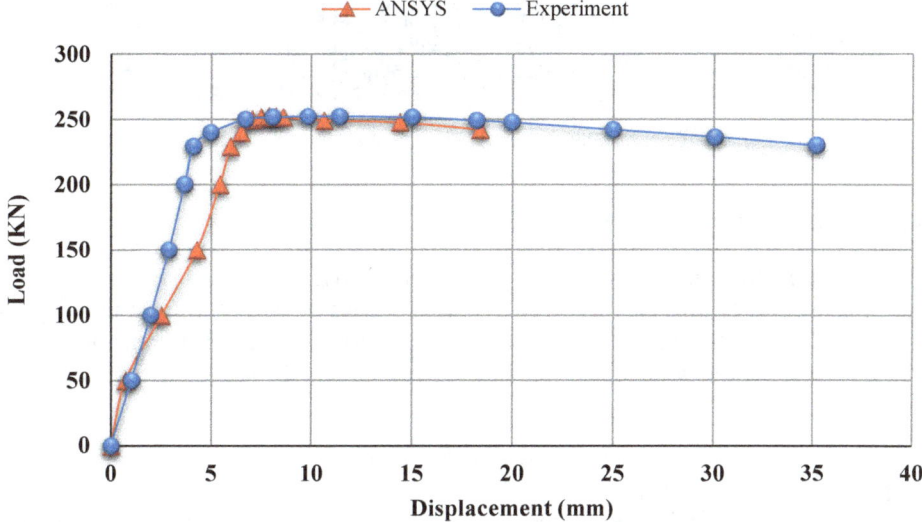

Fig. 8. P-Delta Curve for specimen CRB-B

Crack patterns from Nor's Experiment [1] and from our ANSYS verification models are shown in the following figures (Figs. 9 and 10).

Fig. 9. Crack patterns from Nor's experiment [1]

As we notice, there is good agreement between ANSYS results and the experimental results, which means that we can trust on the analytical models and can predict results from our parametric study.

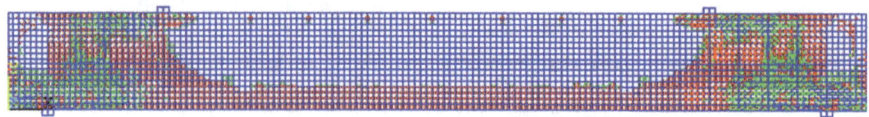

Fig. 10. Crack patterns from the verification models on ANSYS

6.2 Results of the Parametric Study

This section includes the analysis of the results recorded from the Finite Element Models that was previously presented in Sect. 5. The objective of this analysis is to study the effect of the investigated parameters on the behavior of reinforced concrete beams. Table 4 illustrates the failure loads and their corresponding displacement for each beam.

Table 4. Results of the parametric study

Group	Beam no.	Stirrup's position	Type of stirrup	$P_{failure}$ (KN)	$\Delta_{failure}$ (mm)
1	B1	Vertical	Two-legged	310	12.98
	B2	Inclined (45°)	Two-legged	350	16.1
	B3	Combination	Two-legged	320	14.4
2	B4	Vertical	Single-legged·	310	12.35
	B5	Inclined (45°)	Single-legged	330	18.4
	B6	Combination	Single-legged	310	15.11
3	B7	Inclined (30°)	Two-legged	350	13.35
	B8	Inclined (60°)	Two-legged	270	12.64

6.2.1 Effect of the Installation Position for Stirrups

The first parameter in current parametric study is the installation position for stirrups. B1 and B4 have stirrups which installed vertically but the rest of the beams have stirrups that is installed in inclined position or a combination between vertical and inclined. In this parameter we will focus on Group (1) only. Figure 11 illustrates the results.

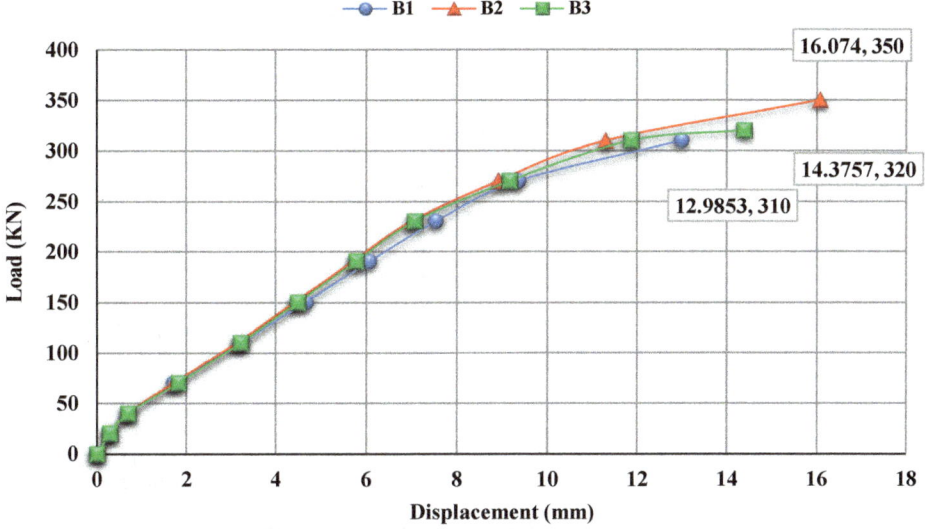

Fig. 11. P-Delta Curves for specimens in Group (1)

The comparison shows that installing the stirrups in inclined position with 45° increases the beam capacity about 12% when all stirrups are inclined along the beam length (B2). But when the inclination occurs at the shear zone only (B3) in order to make a combination between vertical and inclined stirrups, the beam capacity increases with 3% only. The following figures illustrates the crack patterns of beams. Cracks in green color are critical and causes failure, so it is illustrated separately to be clear (Fig. 12).

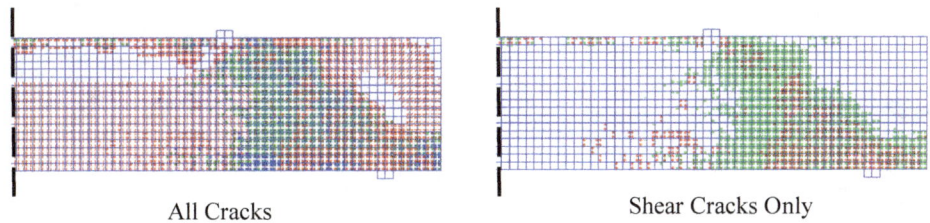

All Cracks   Shear Cracks Only

Fig. 12. Crack patterns

6.2.2 Effect of Using Different Types of Stirrups

The second parameter in current parametric study is the type of stirrups (two-legged and single-legged). All beams in Group (1) and Group (3) have two-legged stirrups, and all beams in Group (2) have single-legged stirrups. In this parameter we will focus on Group (1) and Group (2) only. Figure 13 illustrates the results.

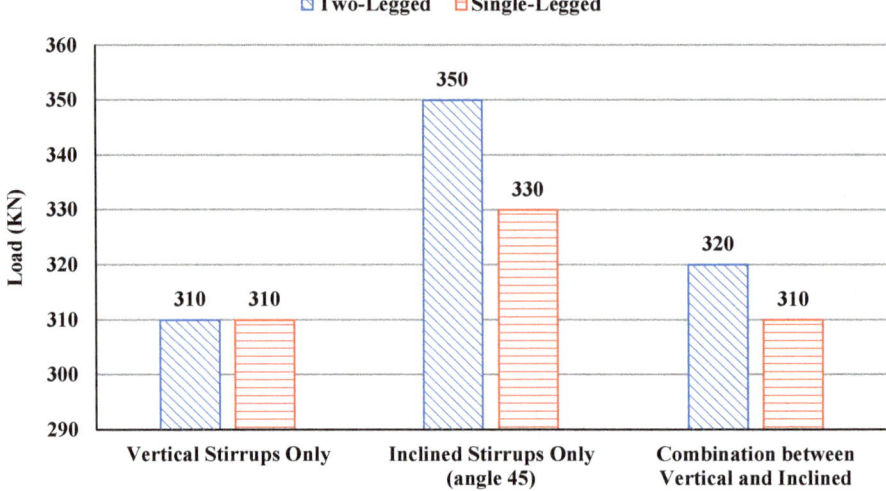

Fig. 13. Failure loads for Group (1) and Group (2)

As illustrated above in Fig. 13, using the two-legged stirrups is more effective than using the single-legged especially in inclined stirrups with 45°. And this is because of confinement that two-legged stirrups make. The following figure illustrates a comparison between their deflections (Fig. 14).

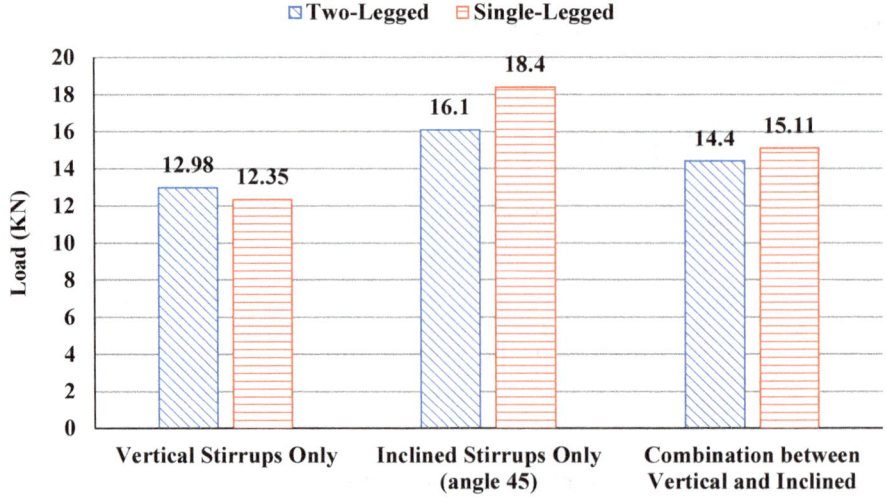

Fig. 14. Deflection for Group (1) and Group (2)

6.2.3 Effect of the Angle of Inclination

The third parameter in current parametric study is the angle of inclination. Three angles of inclination are studied (30–45–60). In this parameter we will focus on Group (3) to study the effect of inclination angle on beams capacities. B2 has angle of 45°, B7 has angle of 30° and B8 has angle of 60°. Figure 15 illustrates the results.

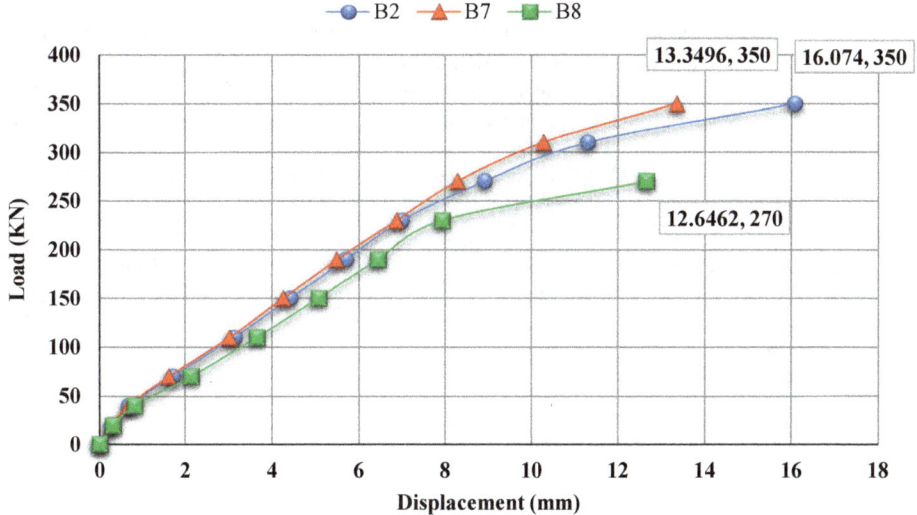

Fig. 15. P-Delta Curves for specimens in Group (3)

The comparison shows that the beams with inclination angles 30° and 45° have the same capacities but the inclination angle 30° enhances the deflection, as it decreases the deflection about 17%. On the other hand, the inclination angle 60° decreases the capacity of beams about 23% when it is compared with the other two angles of inclination 30° and 45°.

7 Conclusions

a) From the verification models, there is good agreement between ANSYS results and previous experimental work.
b) Inclined stirrups are more effective in enhancing the shear capacity than the vertical stirrups. As it increases the capacity about 12%.
c) Using the stirrups as a closed loop (two-legged) causes confinement for concrete that leads to enhance their capacity more than using the single-legged, especially in inclined stirrups.
d) Using angles of inclination 30° and 45° is more effective than using angle 60°. As the inclination angle 60° decreases the capacity of beams about 23% when it is compared with the other two angles of inclination 30° and 45°.

References

1. Zamri, N.F., Mohamed, R.N.: The Effect of Inclined Shear Reinforcement in Reinforced Concrete Beams. Malays. J. Civ. Eng. **30**(1), 85–96 (2018). https://doi.org/10.11113/mjce.v30.169
2. Saravanakumar, P., Govindaraj, A.: Influence of vertical and inclined shear reinforcement on shear cracking behavior in reinforced concrete beams. Int. J. Civ. Eng. Technol. (IJCIET) **7**(6) (2016). ISSN Print: 0976-6308 and ISSN Online: 0976-6316
3. Deepthi, K., et al.: Experimental investigation of shear behavior in flexure members. Int. J. Recent Technol. Eng. (IJRTE) **7**(6C2) (2019). ISSN 2277-3878
4. Patil, S.H., et al.: Increase the shear load carrying capacity of reinforced concrete beam by providing inclined multiple stirrups in shear zone. Int. Res. J. Eng. Technol. (IRJET) (2016). e-ISSN 2395 -0056, p-ISSN 2395-0072
5. Sayyad, A.S., Patankar, S.V.: Effect of stirrups orientation on flexural response of RC deep beams. Am. J. Civ. Eng .Archit. **1**(5), 107–111 (2013). https://doi.org/10.12691/ajcea-1-5-4
6. Kim, H., et al.: Shear behavior of concrete beams reinforced with GFRP shear reinforcement. Int. J. Polym. Sci. **2015**, Article ID 213583, 8 p. (2015). http://dx.doi.org/10.1155/2015/213583
7. Karayannis, C.G., et al.: Shear capacity of RC rectangular beams with continuous spiral transversal reinforcement. WIT Trans. Model. Simul. **41** (2005). www.witpress.com. ISSN 1743-355X (on-line)

Behavior of Reinforced SCC Short Columns Subjected to Weathering Effects

Hassan Ahmed[1(✉)] and Mostafa Yossef[1,2]

[1] Arab Academy for Science, Technology and Maritime Transport, Cairo, Egypt
hassan.ahmed@staff.aast.edu
[2] Aalto University, Espoo, Finland

Abstract. Bridges & offshore structures are often subjected to fresh and saltwater wet-dry cycles, freeze-thaw cycles in winter, and heating cycles in summer. These environmental effects often cause steel reinforcement corrosion which can be protected using zinc coating. Zinc coating acts as a protective layer for the steel against corrosion. In this paper, the behavior of steel bars with and without zinc coating is investigated. An experimental study is performed on short reinforced concrete columns. Short 50-cm length concrete columns with square 10×10-cm cross-section, reinforced by four steel bars of 10-mm diameter are subjected to various weathering effects: accelerated corrosion, cycles of freshwater wet-dry, saltwater wet-dry, Freeze-Thaw (FT) and heating. Self-compacted concrete (SCC) mixes are prepared with 28 MPa average strength. A total of 20 concrete column samples are casted and divided into five groups: 1) salt water-wet-dry-uncoated rebars, 2) salt water-wet-dry-zinc-coated rebars, 3) fresh water-wet-dry-uncoated rebars, 4) heating to 70 °C, and 5) freeze-thaw samples, where 5 samples acted as a control sample and another 5 samples are subjected to accelerated corrosion. The remaining 10 samples are divided among the five groups where each group has 2 samples that are subjected to deterioration cycles. Results indicated that samples subjected to saltwater wet-dry cycles had the least strength among the five groups as expected. However, the diameter and mass losses in accelerated corrosion samples are doubled the values of that of the saltwater dry-wet cycles with the same mean strength.

Keywords: Self-compacted concrete · Accelerated corrosion · Freeze-Thaw · Wet-dry · Zinc coated-rebars

1 Introduction

Sustainable urban cities are one of the main goals that the world is trying hard to achieve nowadays. Today, 55% of the world population is living in urban cities and it is estimated to rise to 68% by 2050 (United Nations 2019). With increasing number of the population in urban areas, the construction of new buildings and structures in addition to maintenance of older structures become an urgent need. The concrete industry has the highest share in the Middle East region due to its relatively cheap cost compared to steel and timber. However, while reinforced concrete is a good thermal insulation material, it

© The Author(s), under exclusive license to Springer Nature Switzerland AG 2021
H. Shehata and S. El-Badawy (Eds.): *Sustainable Issues in Infrastructure Engineering*, SUCI, pp. 93–106, 2021.
https://doi.org/10.1007/978-3-030-62586-3_7

often fails to resist other deterioration effects mainly due to corrosion of reinforcement steel and chloride effects on the concrete material.

Deterioration of concrete can be likely seen in bridges in reinforced concrete columns subjected to either fresh or saltwater wet-dry cycles or dry columns subjected to cycles of freeze and thaw in cold areas or heating and cooling down in hot regions. Recent research had studied the effect of weathering effects by replacing the weathering cycles with accelerated corrosion to achieve faster results (Sanz et al. 2018; Ye et al. 2018). However, there's a need to investigate the validity of the accelerated corrosion to replace different weathering effects. In this paper, short reinforced concrete columns, cast using self-compacted concrete (SCC), were subjected to various weathering effects and compared with accelerated corrosion results.

SCCs were developed to provide the construction industry with more durable concrete that can resist the weathering effects. This can be obtained through the high workability of this concrete which could help with casting small sections, heavy-reinforced sections, repaired members, or members where it will be critical to use internal vibrators for compaction. Concrete compacted without the aid of vibrators, increases the bond between concrete and the embedded reinforcement, which improves the performance of the cast members, also, the low permeability of this high workable mix will lead to a more durable structure. SCC, due to its high mobility, can fill the formwork without any mechanical help, and with no risk of segregation (Paultre et al. 2005).

Some researchers dealt with SCC through the mix proportioning, and properties of concrete in the fresh and hardened stages (Sharifi 2012). Others studied the structural performance of such concrete. When studying the flexural and shear behavior of SCC reinforced beams versus the normal reinforced concrete beams, Luo and Zheng (2005) found that there is nearly no difference in performance between the two concretes regarding the failure mechanism, the yielding and ultimate moments, and the shear capacity. Therefore, SCC does not affect the structural behavior of members, however, it may have a positive effect on structures' durability.

The low-cost versatile material, concrete, must withstand the attack of the surrounding deterioration factors to survive in severe environments. These factors include, but not limited to, attack of chlorides, sulphates, carbonation, freezing-thawing, heating to elevated temperatures, drying-wetting cycles in tidal zones, etc. According to Wang and Li (2014), concrete is a heterogeneous material with passageways for water and aggressive ions penetrations. This phenomenon decreases the probability of surviving in severe environments if durability is not taken into consideration when designing concrete members. Current studies are still searching behind the environmental effects on concrete and how to protect concrete mainly from the corrosion of its reinforcement. Attack of saltwater and carbonation may be considered the main factors for reinforcement corrosion. Ghanooni-Bagha et al. (2020) stated that slight adjustments in concrete proportioning, water-cement ratio and cement content, may delay the corrosion initiation and improve concrete durability.

2   Experimental Study

An experimental study is performed on self-compacted short reinforced concrete columns. Short 500 mm length concrete columns with square 100×100 mm cross-section, reinforced by four steel bars of 10 mm diameter are subjected to various weathering effects such as cycles of freshwater wet-dry, saltwater wet-dry, Freeze-Thaw (FT) and heating and cooling and compared with accelerated corrosion. A total of 20 concrete column samples are cast and divided into five groups: 1) salt water-wet-dry-uncoated rebars, 2) salt water-wet-dry-zinc-coated rebars, 3) fresh water-wet-dry-uncoated rebars, 4) heating to 70 °C and cooling down to room temperature, and 5) freeze-thaw samples, where 5 samples act as control samples and another 5 samples are subjected to accelerated corrosion as shown in Fig. 1. The remaining 10 samples are divided among the five groups where each group has 2 samples that are subjected to deterioration cycles.

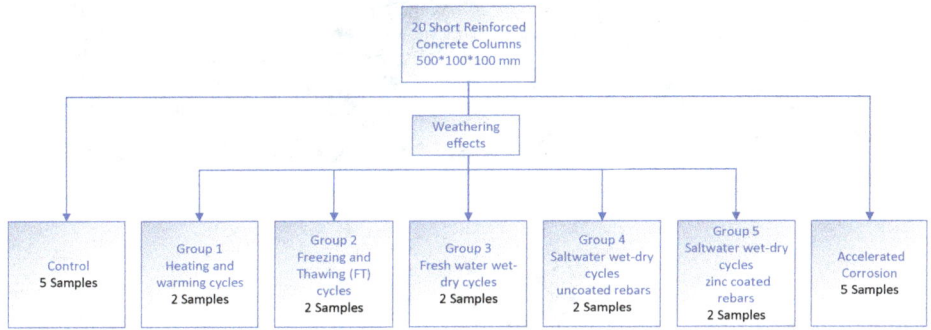

Fig. 1. Experimental program

2.1   Sample Preparation

Self-compacted concrete mix design was chosen as shown in Table 1 to facilitate the casting of the concrete in relatively small samples. Portland cement, sand, coarse aggregate, and water were all mixed in 0.142 m^3 concrete mixer with the addition of superplasticizer to facilitate the concrete workability. Slump test was recorded at 17.3 cm showing good workability of the mix design.

Table 1. Mix design for self-compacted concrete

Cement (kg/m^3)	Sand (kg/m^3)	Coarse aggregate (kg/m^3)	Super plasticizer (kg/m^3)	Water (kg/m^3)
350	945	850	8.8	210

For steel reinforcement, 10 mm diameter deformed steel rebars were cut into three 470 mm long rebars and one 550 mm long rebar. The four rebars are held together using four stirrups of 6 mm diameter where each side is 70 mm width to fit inside the 100 × 100 mm formwork as shown in Fig. 2. Five steel cages were cleaned from dust using a cleaning agent and coated with Sika Zinc Rich® coating using hand brush as shown in Fig. 3 where two sides were coated first then left to dry for two hours, then the other two sides were coated and left to dry. While the most effective way to avoid corrosion is to apply the coating before installing the stirrups, the samples in this study were coated after the installation of the stirrups to simulate actual site conditions.

Fig. 2. Reinforcement steel cage

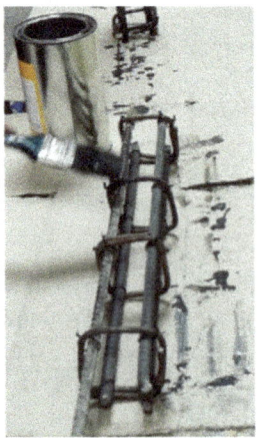

Fig. 3. Applying zinc coating to steel reinforcement cages

Plywood formworks were prepared to fit reinforced concrete columns with a cross-section of 100 × 100 mm and a length of 500 mm. The concrete cover was 15 mm from each side so that steel cage cross-section including stirrups was 70 × 70 mm. The steel cage was placed first so that it fits inside the formwork except for the 550 mm long rebar which was left out to be used for accelerated corrosion's electrical connections. A 150 × 150 × 150 mm cubic steel formwork was prepared for casting to evaluate the strength of the concrete after 28 days. The concrete was cast five times into 9 formworks of columns and cubes at a time for each group as shown in Fig. 4, however, only 4 samples per patch were used for the current study. After curing for 28 days, the 550 mm rebar and the upper surface of three columns samples out of four samples per each group were cut and ground, respectively. The remaining sample of each group with overhanging rebar was used in the accelerated corrosion process.

Fig. 4. Casted reinforcement concrete (RC) samples

2.2 Accelerated Corrosion

Five concrete columns were placed in a container filled with saltwater up to two-third of the column's height. The salt percentage was set to 3.5% of the total volume of the water to simulate the average seawater salt percentage. A rebar was placed in the water to act as the cathodic element while the reinforcement bars inside the columns act as the anode elements. The external driven voltage was set to 12 V and connected in parallel to all five samples as shown in Fig. 5. The saltwater was changed once throughout the total duration of 17 days and as the dark brown rust started accumulating on the surface as shown in Fig. 6, the accelerated corrosion stopped, and the columns were then cleaned for compressive testing.

Fig. 5. Connected circuit for accelerated corrosion

Fig. 6. Accumulated rust on the surface of the columns after accelerated corrosion

2.3 Weathering Effects

Two samples of each group were subjected to different conditions to simulate actual weathering cycles. These cycles are divided into five groups as follows: 1) salt water-wet-dry-uncoated rebars, 2) salt water-wet-dry-zinc-coated rebars, 3) fresh water-wet-dry-uncoated rebars, 4) heating to 70 °C and cooling down to room temperature, and 5) freeze-thaw samples.

The first and the second group consisted of reinforced concrete samples without and with Zinc-coated rebars, respectively. The two samples were placed in a large ceramic tiled sink and subjected to saltwater with different water levels. The salt percentage was set to 3.5% which is the same of accelerated corrosion setup for comparison purpose while the water level was set to one-third (1/3) of the column height for 4 days and raised to two-thirds (2/3) of the column height for 3 days to simulate the water tide as shown in Fig. 7. A total of 11 cycles over three months were performed before the samples were left to dry in the open air for a week before testing. A similar procedure was applied to the third group but with fresh tap water.

Fig. 7. Samples subjected to saltwater wet-dry cycles with water level is set to (a) 2/3 of column height for 3 days and 1/3 of column height for 4 days

The fourth group was subjected to freezing in a deep freezer with the temperature setting of −2 °C for 7 days as shown in Fig. 8 and then left outside the freezer for 4 days for thawing in average laboratory temperature of 20 °C. This cycle was repeated for 6 cycles. The −2 °C temperature was chosen to simulate average low temperature in the Middle East region and specifically Egypt where the low temperature occurs at high altitude mountainous areas. The fifth group was heating up to 70 °C in the oven for 1 day and cooled down to room temperature as shown in Fig. 9. A total of 50 cycles were performed over a total period of 100 days.

Fig. 8. Samples placed on a deep freezer and subjected to −2 °C freezing and thawing cycles

Fig. 9. Samples placed on an oven and subjected to 70 °C heating cycles

3   Test Results and Discussion

Initial material testing was performed to determine the material properties used. Self-compact concrete cubes were tested under compressive testing using ELE compression test machine. The mean value of the compressive strength of the concrete cubes was 28 MPa. In addition to concrete, samples of 10 mm diameter steel bars were tested under tension loading using an Instron universal machine where a typical load-elongation curve is shown in Fig. 10.

Instron universal machine was used to determine the load capacity of the columns by compressive loading of the samples under displacement control. 5 samples were tested as control samples as discussed earlier in Sect. 2 and Fig. 1. Typical compressive testing of concrete columns is shown in Fig. 11(a). First, the upper concrete surface suffered from concrete spalling as shown in Fig. 11(b), then the concrete column held the load until it finally failed showing longitudinal cracks along the reinforcement and concrete spalling at the bottom of the concrete column as shown in Fig. 11(c). Load-contraction curves of the control samples are shown in Fig. 12.

Figure 14(a) shows compression testing of accelerated corroded concrete columns samples. Load-contraction results for the tested samples are displayed in Fig. 13. Spalling of the cover were the typical failure mode of the tested samples as shown in Fig. 14(b). After concrete columns testing, the steel rebars were extracted, cleaned thoroughly using steel wire mesh, cleaning agent and grinding paper to remove any rust then, the diameter and mass losses of the steel bars were measured. A similar process was repeated for all different group to determine the mass, diameter and load capacity losses as shown in Fig. 15.

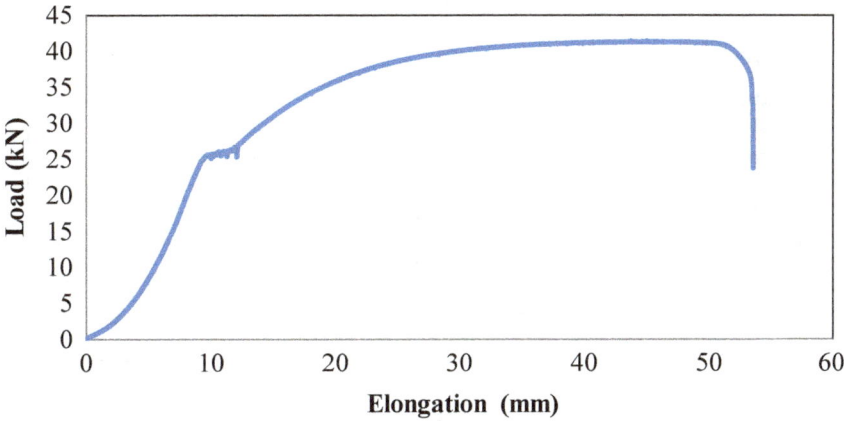

Fig. 10. Typical load-elongation curve for reinforcement steel

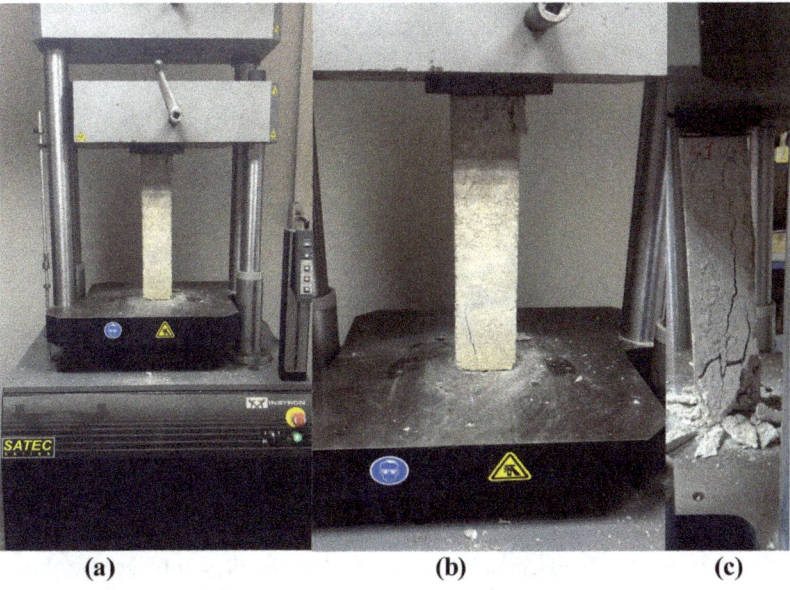

Fig. 11. (a) Compression testing of control RC columns (b) initial failure mode through spalling of the upper surface (c) final failure modes showing longitudinal cracks and lower surface spalling

Figure 15 shows that all weathering effects have limited diameter losses ranges from a mean value of 0.87% for heating and cooling cycles to 2.27% for freezing and thawing cycles. Mass losses were following the same trend for diameter losses however with constant double the value of the diameter losses which indicates that both indicators can lead to the same conclusion. In the other hand, the average load capacity for different samples experienced larger values of losses compared with diameter and mass losses.

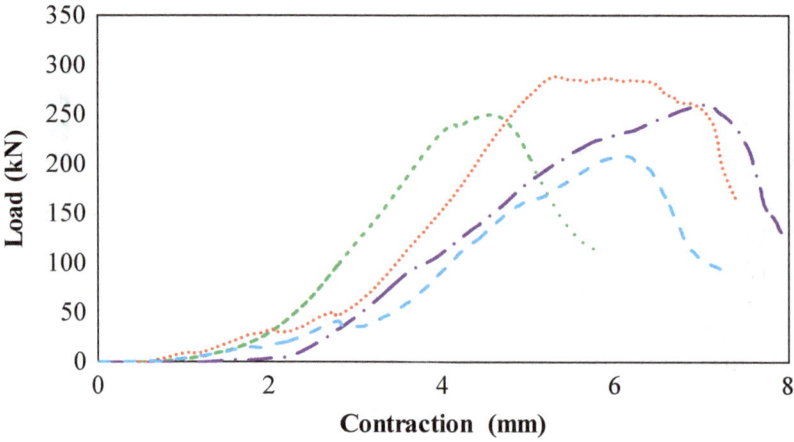

Fig. 12. Load-contraction curves for control RC columns

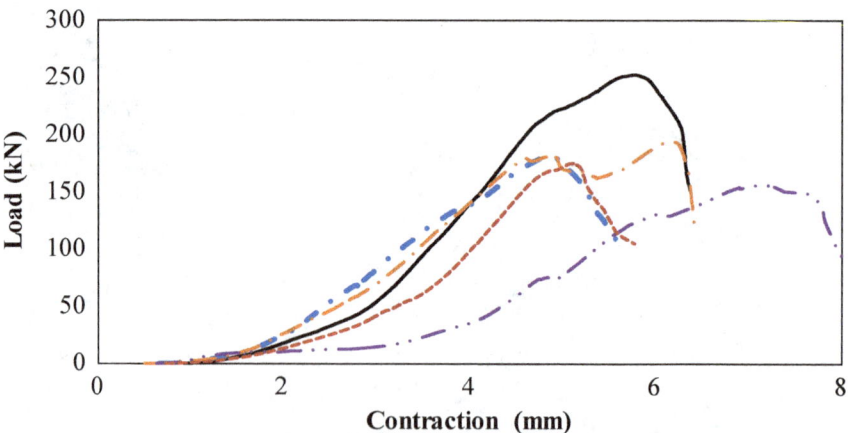

Fig. 13. Load-contraction curve for RC columns subjected to accelerated corrosion

The lowest average load capacity was 13% for samples subjected to heating and cooling cycles, followed by 17% for samples subjected to freezing and thawing cycles, then, 25% for samples subjected to fresh/tap water dry-wet cycles. The highest average load capacity loss was 33% for samples subjected to saltwater wet-dry cycles with coated and uncoated rebars. This can indicate that wet-dry cycles have a limited effect on the corrosion of the steel rebars.

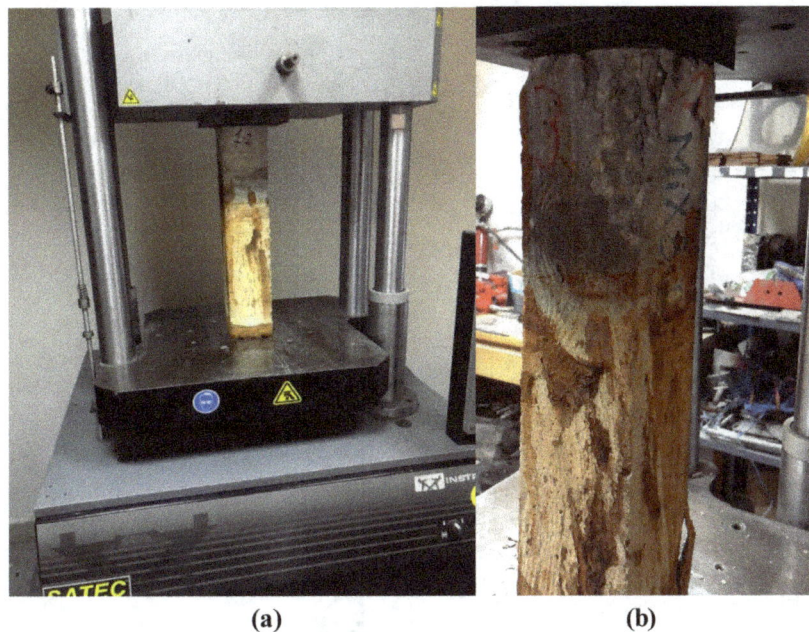

(a)   (b)

Fig. 14. (a) Compression testing of accelerated corroded RC columns (b) close caption showing spalling of the concrete cover

Figure 15 can also show a clear comparison between the effects of samples subjected to accelerated corrosion and saltwater wet-dry cycles. Samples with uncoated rebars experienced a close value for both accelerated corrosion and salt wet-dry cycles of an average load capacity loss of 30% and 33%, respectively. However, the diameter and mass losses values for the accelerated corrosion were almost three times that of the saltwater wet-dry cycles raising from 1.9% and 3.7% to 6% and 13% for mass and diameter losses, respectively. That indicates that saltwater wet-dry cycles affect the concrete properties more than the corrosion of the reinforcement steel.

The effect of the zinc coating can be observed through the low losses of load capacity of 13% for coated samples compared to 30% for uncoated samples when subjected to accelerated corrosion. However, the zinc coating has a limited effect when the samples subjected to saltwater wet-dry effect.

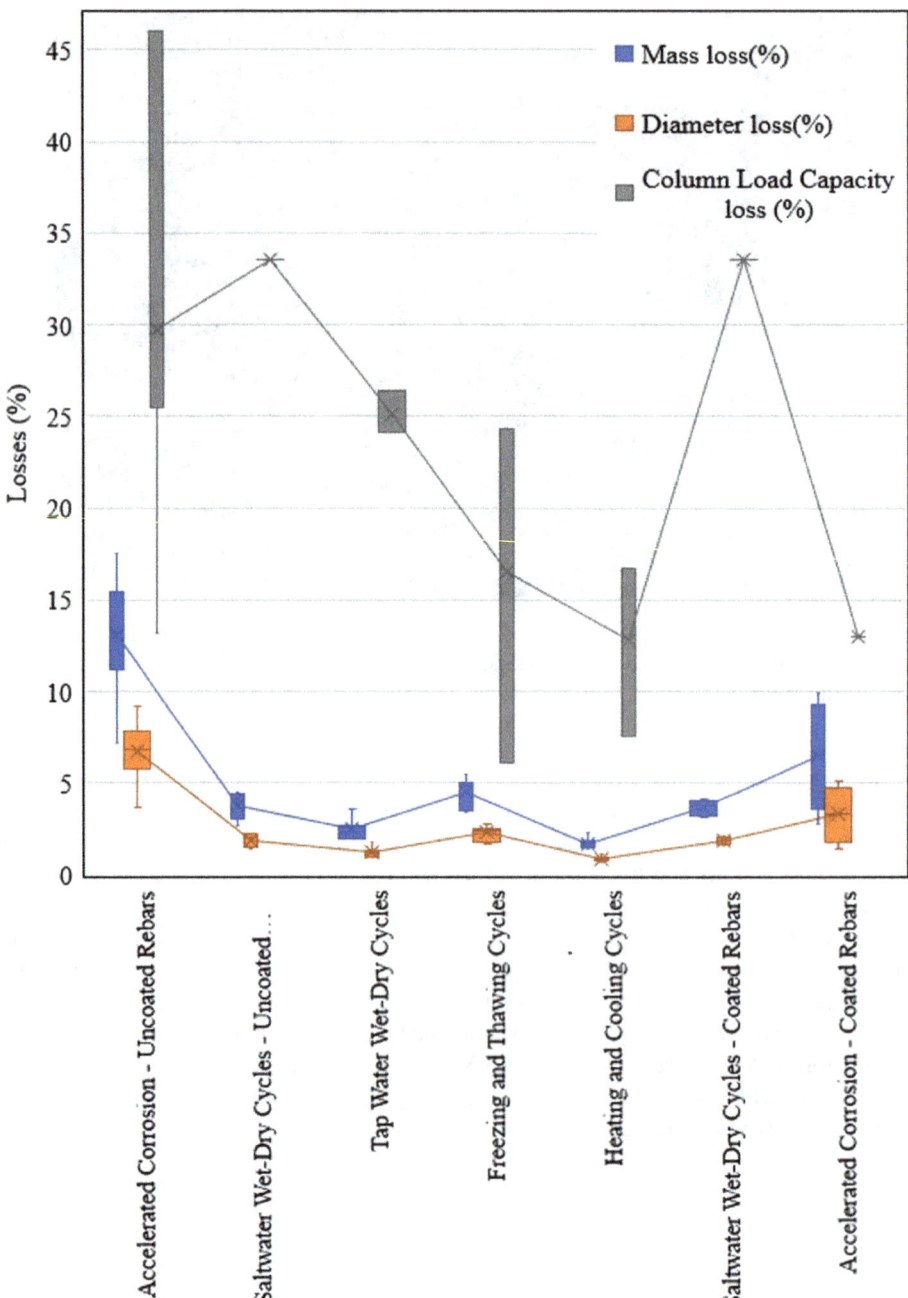

Fig. 15. Losses for samples subjected to different effects

4   Conclusions and Future Recommendation

Steel reinforced concrete-short columns are subjected to different weathering conditions such as saltwater and fresh wet-dry cycles, freezing and thawing, and heating and cooling. The results were compared to control samples and samples subjected to accelerated corrosion samples. Through this study, several conclusions can be noticed as follows:

1. All weathering effects have limited diameter losses ranges from a mean value of 0.87% for heating and cooling cycles to 2.27% for freezing and thawing cycles.
2. Mass losses were following the same trend for diameter losses however with constant double the value of the diameter losses ranging from 1.73% for heating and cooling cycles to 4.52% for freezing and thawing cycles.
3. The average load capacity for different samples experienced larger values of losses compared with diameter and mass losses ranging from 13% for heating and thawing cycles to 33% for samples subjected to saltwater wet-dry cycles with coated and uncoated rebars.
4. Uncoated rebar columns subjected to saltwater wet-dry cycles have similar performance to similar samples subjected to accelerated corrosion regarding load capacity, however, the diameter and mass losses values for the accelerated corrosion were almost three times that of the saltwater wet-dry cycles raising from 1.9% and 3.7% to 6% and 13% for mass and diameter losses, respectively. That indicated that saltwater wet-dry cycles affect the concrete properties more than the corrosion of the reinforcement steel.
5. Zinc coating of rebars has a limited effect when the samples subjected to saltwater wet-dry effect while it effective when samples subjected to accelerated corrosion.

Therefore, we can conclude that accelerated corrosion can only simulate the corrosion without the overall degradation induced from the wet-dry cycles. Ongoing research is currently carried by the authors to investigate the effect of wet-dry cycles on concrete without steel corrosion.

Acknowledgments. The authors would like to thank the material laboratory technician Mr. Soliman Mohey and former undergraduate students; Omar Nabil, Mohamed Rizk, Ahmed Amr, Monzier Mahmoud and Mohamed Magdy for assisting in manufacturing and testing of the samples. The second author would like to acknowledge the support of Aalto University (Postdoc/T214).

References

Ghanooni-Bagha, M., YekeFallah, M.R., Shayanfar, M.A.: Durability of RC structures against carbonation-induced corrosion under the impact of climate change. KSCE J. Civ. Eng. **24**(1), 131–142 (2019). https://doi.org/10.1007/s12205-020-0793-8

Luo, S., Zheng, J.: Research on bending and shear behavior of self-consolidating concrete beams. In: Proceedings of First International Symposium on Design, Performance and Use of Self-Consolidating Concrete, SCC'2005-China, Changsha, Hunan, China (2005)

Paultre, P., Khayat, K.H., Cusson, D., Tremblay, S.: Structural performance of self-consolidating concrete used in confined concrete columns. ACI Struct. J. **102**, 560–568 (2005)

Sanz, B., Planas, J., Sancho, J.M.: Study of the loss of bond in reinforced concrete specimens with accelerated corrosion by means of push-out tests. Constr. Build. Mater. **160**, 598–609 (2018)

Sharifi, Y.: Structural performance of self-consolidating concrete used in reinforced concrete beams. KSCE J. Civ. Eng. **16**, 618–626 (2012)

United Nations: World Urbanization Prospects: The 2018 Revision. Department of Economic and Social Affairs, Population Division (2019)

Wang, L., Li, S.: Capillary absorption of concrete after mechanical loading. Mag. Concr. Res. **66**, 420–431 (2014)

Ye, H., Fu, C., Jin, N., Jin, X.: Performance of reinforced concrete beams corroded under sustained service loads: a comparative study of two accelerated corrosion techniques. Constr. Build. Mater. **162**, 286–297 (2018)

Implications of Different Foamed Bitumen Stabilization Production and Curing Processes on Airport Pavement Thickness and Life

Greg White[(✉)] and Tom Weir

University of the Sunshine Coast, Sippy Downs, QLD, Australia
gwhite2@usc.edu.au

Abstract. Foamed bitumen stabilization is a useful and well established method for improving crushed rock and natural gravel materials for pavement construction. Like most pavements, those with a foamed bitumen base course (FBB) are usually designed using layered elastic softwares in which the FBB layer is characterized by an elastic modulus and a Poisson's ratio, with the modulus having a significant influence on pavement thickness. In Australia, FBB characterization is based on the saturated indirect tensile modulus after three days of accelerated curing of samples produced in a laboratory mixer. It is well established that this approach to FBB characterization is not representative of field production and in-pavement curing conditions. To determine the effect of FBB production and curing on FBB modulus, pavement thickness and predicted pavement life, the same FBB was produced using a laboratory mixer, an exsitu pugmil and an insitu stabilizer. Material was sampled and the uncured, cured and saturated modulus was measured after various periods of accelerated laboratory curing. The exsitu produced FBB was also cured using simulated in-pavement conditions. The various FBB modulus values were then used to determine the required thickness and predicted life of a typical aircraft pavement including a 300 mm thick FBB layer. It was found that field produced FBB modulus increased significantly during the first 90 days after production and that laboratory production and curing protocols were not representative of field production and in-pavement curing conditions. Layered elastic pavement modelling showed that more than 80% of the predicted pavement damage occurred in the first 20 days after FBB production. It is therefore recommended that FBB remains untrafficked for 7–12 days after production, wherever possible, and thinner pavements are likely to perform adequately in situations where the FBB is protected from traffic loading for more than 14 days following production.

1 Introduction

Foamed bitumen stabilization (FBS) is the process of mixing foamed bitumen into a granular material to stabilize or improve the material properties (Austroads 2018). The bitumen is foamed by combining hot bitumen and cool water, explosively vaporizing the water and temporarily increasing the surface area to volume ratio of the bitumen. When the stabilized material is used as a base course material in an existing or a new pavement

H. Shehata and S. El-Badawy (Eds.): *Sustainable Issues in Infrastructure Engineering*, SUCI, pp. 107–126, 2021.
https://doi.org/10.1007/978-3-030-62586-3_8

structure, it is commonly referred to as foamed bitumen base (FBB). As detailed below, FBB can be produced from new materials or existing pavement materials, from crushed rock (CR) or from natural gravel (NG) and the parent material can be first-class or marginal in nature (Austroads 2018).

FBS provides attractive properties that make it a valuable option in many pavement construction and rehabilitation applications. First, it has a significantly higher modulus than CR, meaning that a given thickness of FBB provides more structural contribution to a pavement than the same thickness of CR (Austroads 2019). Second, it is generally moisture resistant, providing resilient infrastructure in areas that are prone to flooding and a high water table (White 2014). Third, it is rapid and easy to construct, not needing to dry-back (like CR), or to cool (like asphalt), or set (like cement treated base) before being covered (White 2018). Furthermore, layers of FBB bond well to other layers of FBB without a prime or tack coat. Finally, FBB is not prone to cracking like cement treated base (White et al. 2018). These properties of FBB are particularly useful for expedient airport pavement rehabilitation works (White 2017).

Most airports have only one main runway and single runways can not usually be closed for extended periods of time to allow rehabilitation works to be complete. As a result, airport pavement rehabilitations are usually performed during short night-time closures and the runway reopened to aircraft traffic each day (AAA 2017). These expedient rehabilitation works are expensive and represent significant risk and any opportunity to increase their efficiency is valuable. FBB offers reduced rehabilitation depth (due to the higher modulus) and is constructed efficiently (with thick layers, good self-bonding and rapid trafficability) making it attractive for airport applications. The only significant disadvantage associated with FBB is the relatively high cost, compared to other stabilisation options (Austroads 2019).

In Australia, FBB has been used to rehabilitate or otherwise upgrade pavements at Saint George, Melbourne, Darwin (White 2014), Carnarvon, MacArthur River, Barimunya, Gladstone (White 2018), Brisbane (Soufi-Sabbagh 2018) and Whitsunday Coast airports (White et al. 2018) since 2010. The applications have included marginal material improvement, expedient strengthening and moisture sensitivity reduction. In all these cases, layered elastic software was used to determine the thickness of the pavement required to accommodate the predicted aircraft traffic loads over the design life of the pavement. As with all layered elastic softwares for pavement design, the FBS materials were characterized by an elastic modulus and a Poisson's ratio, with the modulus having a greater influence on the required pavement thickness (Huang 1993).

The selection of a representative elastic modulus of each material is critical to the reliability and efficiency of pavement thickness design. Setting the representative modulus too high will result in an understrength pavement that may fail prematurely, while setting the modulus too low will introduce unnecessary conservatism. Excessive conservatism is particularly concerning for expedient pavements because the increase in thickness associated with a more conservative pavement design can lead to expedient construction challenges and will reduce the quality of construction, introducing the risk of poor pavement performance (White 2017). Therefore, it is critical that the thickness design of expediently constructed pavements be based on realistic modulus values.

The modulus of FBS materials is generally either assumed for thickness design purposes and then the material is designed to just exceed this value, or the modulus of the proposed FBS material is measured in the laboratory and a characteristic value is used to design the pavement structure and thickness. Either way, it is important that the laboratory measured modulus is representative of that expected in the field. However, recent comparisons of laboratory mixed and cured samples, with nominally identical samples produced in the field and cured within a pavement, indicate significant differences between Australia's standard laboratory conditions and field curing (Austroads 2018). This discrepancy must be addressed to ensure that pavements that include significant thicknesses of FBB take appropriate account of the increased structural contribution associated with FBS.

The aim of this research was to determine the difference in modulus associated with field and laboratory produced and cured FBB and how these differences impact typical pavement thicknesses and predicted structural life. A typical CR was stabilized and the FBB modulus was measured after different curing periods, both in the laboratory and using a simulated in-pavement curing condition. The different modulus values were used to design a typical FBS airport pavement for typical aircraft traffic and the effect of production process, curing condition and curing time were all considered.

2 Background

As stated above, both CR and NG can be improved by FBS. In general, the improvement will be greater for a NG but the actual performance of the FBS material will generally be better for CR than for NG. As discussed below, each can be produced using insitu or exsitu processes and both are characterized using a laboratory mixer and standardized laboratory curing conditions.

2.1 Foamed Bitumen Base Production

FBB can be produced in a pugmil (exsitu) or using a pulverizing and stabilising machine (insitu) (Fig. 1). It is logical that insitu stabilization is suited to the reuse of existing materials (CR or NG) and exsitu processing is most appropriate when new CR is used to produce FBB. Insitu construction of FBB reduces the cost associated with double handling of existing materials, as well as the mobilization of separate production and paving equipment. As a result, insitu processing is more cost effective than exsitu production when existing pavement materials are being reused. However, exsitu processing provides the opportunity to inspect and assess the exposed subgrade, test the excavated existing material and perform any necessary secondary pulverization of any existing asphalt surface. Equipment break-down risk is also reduced by producing the full volume of FBB required for the work period, prior to the commencement of excavation. Furthermore, it is expected that exsitu production in a pugmil produces a material that has less moisture and bitumen content variability than insitu processing. In contrast, multiple passes of insitu stabilizers are expected to pulverize existing pavement materials better than typical excavation equipment, such as cold planing machines.

Fig. 1. Typical (a) insitu (b) exsitu and (c) laboratory FBS mixing equipment

Regardless of the production process, FBS mixture design and characterization is generally performed in a laboratory mixing device (Fig. 1). Exsitu, insitu and laboratory FBS processes all use different mixing chambers and mixing times, resulting in differently mixed materials (Table 1). Aggregate breakdown during stabilization, which changes the overall grading, and bitumen distribution, are both affected by mixing processes and times, meaning the three production processes can result in slightly different materials, even when the mixture design is nominally identical. An example is shown in Fig. 2, which shows visually different samples of nominally identical FBB produced by the different methods. The longer mixing time associated with the laboratory mixer resulted in more uniform distribution of the foamed bitumen, particularly compared to insitu mixing.

Table 1. Comparison of production method mixing (Weir 2020)

Production method	Mixing time (s)	Mixer type
Exsitu	8–18	Pulverising teeth
Insitu	21–51	Lifting paddles
Laboratory	Nominal 360	Count-rotating paddles

360 s mixing is the Australian standard (Austroads 2019) although other jurisdictions use different mixing times

Fig. 2. Example of nominally identical FBB produced (a) insitu, (b) exsitu and (c) in a laboratory mixer

2.2 Foamed Bitumen Base Curing

In Australia, the standard accelerated laboratory curing for FBS materials is unsealed samples placed in a dry 40 °C oven. The cured samples are then tested after conditioning to 25 °C to represent different periods in the pavement lifecycle (Austroads 2018):

- Before curing. Indicative of initial deformation resistance.
- Three days curing. Used for design characterization.
- Seven days curing. Intended to reflect long-term in-field curing.
- Fourteen days during. Intended to reflect very long-term in-field curing.

Testing on the cured samples is performed at the cured moisture content, which is generally low. Saturated testing is performed after vacuum soaking for 30 min and then conditioning to a saturated surface dry state (Austroads 2017). Although standard laboratory curing is intended to represent field curing, but in an accelerated manner, it is clear that field curing conditions vary significantly from site to site and are different from standard laboratory conditions. Furthermore, it is well established that FBB modulus increases significantly during the first three to seven days of accelerated curing and then stabilizes (Austroads 2019). This creates a challenge when developing accelerated laboratory curing and testing protocols because the expected modulus changes over time, both in the laboratory and in the field.

The pavement type, prevailing weather, depth within the pavement, aggregate properties, proximity to the pavement shoulder, soil suction and the height of water table all affect the actual in-pavement curing conditions, with Muller (2017) and Morgan (1972) suggesting that in-pavement moisture conditions vary from 43% of optimum (OMC) to greater than the OMC. Furthermore, it is well established that the curing moisture condition of FBS materials significantly affects the resulting modulus (Austroads 2018; Fu et al. 2010; Wirtgen 2012). Consequently, it has been found that in-pavement cured materials achieve much greater long-term modulus values than those predicted using standard laboratory curing protocols (Austroads 2018). To ensure that realistic modulus values are used for pavement thickness design, it is critical that laboratory curing regimes be representative of field conditions. However, the practical need to accelerate laboratory curing, as well as the significant change in FBB material modulus with time, combine to create a significant challenge.

2.3 Foamed Bitumen Base Characterization

The standard Australian protocols for FBS material characterization for pavement design purposes is based on repeated load, indirect tensile modulus, also known as resilient modulus, measured on samples at 25 °C (Austroads 2019). The FBS material samples used for characterization are those cured in the laboratory for three days and then saturated. In most cases, a standard modulus of 1,500 MPa is used for pavement thickness design and subsequent mixture design adjusts the bitumen and other additives, usually 0.5–1.0% of hydrated lime or cement, to achieve the 1,500 MPa modulus. In some cases, the cost of the bitumen and lime/cement is balanced against the required pavement thickness to determine an optimal solution, with the three day accelerated cured and saturated modulus subsequently used for pavement thickness design. Regardless of the approach taken, there is an important link between the FBB modulus used for pavement thickness design and the three day cured and saturated modulus tested in the laboratory after three days of accelerated curing.

2.4   Airport Pavement Structural Design System

As stated above, almost all modern aircraft pavement thicknesses are determined using layered elastic softwares. Different jurisdictions use different softwares, depending on their local preference. Airport Pavement Structural Design System (APSDS) is preferred in Australia and it is based on the layered elastic program for road pavement design known as CIRCLY (Wardle 1977). APSDS is both transparent and offers great flexibility to the user, and has been calibrated against the FAA's COMFAA (Wardle and Rodway 2010). APSDS enables users to access the full advantages of the layered elastic method, including treatment of aircraft wander across the pavement width and non-standard material parameters, to quickly produce pavement thicknesses for any combination of aircraft and for layered structures or pavements containing non-standard materials. Consequently, APSDS is well suited to parametric analyses, such as sensitivity studies (White 2005), stochastic approaches to pavement thickness determination (White 2009) and evaluation of non-standard materials and design cases (White et al. 2020). APSDS was used in this research, primarily because it allows any FBB characteristic modulus value and pavement composition to be evaluated (AAA 2017).

3   Methods and Results

This research included the measurement of FBB modulus for samples produced to a nominally identical mixture design, but produced using a laboratory mixer, an exsitu pugmil and an insitu stabilizer. Further testing of exsitu produced FBB measured the modulus after accelerated laboratory curing and after simulated in-pavement curing. The various modulus values were then used to design typical FBS aircraft pavements using APSDS, to determine the effect of the different modulus values on pavement thickness and life. Finally, realistic modulus values were estimated at different FBB ages and the values used to compare the relative damage caused to a pavement that was designed using the standard saturated modulus after three days of accelerated laboratory curing as the characteristic FBB modulus value. Where applicable, the results were compared using Student-t tests for the difference of means between populations, with a p-value of less than 0.05 indicating a significant difference in the populations.

3.1   Methods

The FBB produced and tested was a Type 2.3 (TMR 2019) CR from Bromelton quarry in south-east Queensland (Australia) stabilized with 3% foamed bitumen and 1.5% hydrated lime, both by mass of the CR, which is a typical FBB mix design in Australia (Australia 2018). The CR is known to be well suited to FBS, with the pre-stabilized

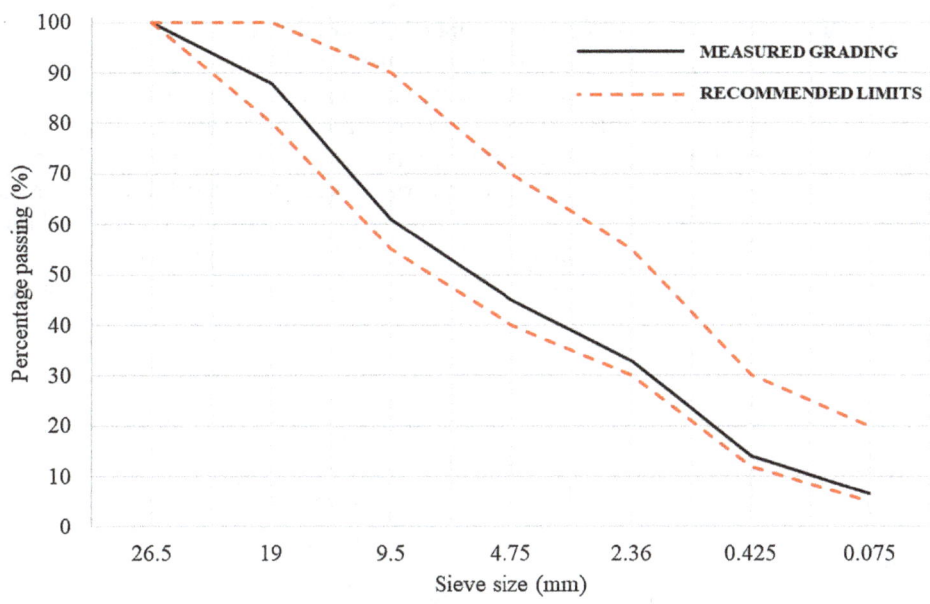

Fig. 3. Pre-stabilized CR grading and recommended grading envelope

grading meeting the recommended envelope (Fig. 3) as fully described and characterised by Weir (2020).

The samples were prepared and tested in accordance with methods used in Australia (TMR 2020), including mixing (TMR Q138), compaction (TMR Q138), curing (TMR Q138), conditioning (TMR Q139) and indirect tensile modulus testing (TMR Q139), with necessary minor modifications for the samples cured in the simulated pavement condition, as detailed below. The material produced exsitu was sampled from a Wirt-gen pugmil and was transported to the laboratory and compacted within four hours of production. The insitu produced material was similarly sampled in bulk from the field, transported to the laboratory and compacted within four hours. Testing of laboratory cured samples was performed prior to oven curing and after three, seven and 14 days of accelerated curing at 40 °C. Pre-cured samples were tested only in the un-saturated condition and the pre-cured materials were tested in duplicate, while the post-curing materials were tested in triplicate, in both cured and saturated conditions.

The samples cured in a simulated pavement were compacted in the laboratory and then placed in an air tight container filled with aggregate conditioned to 4.8% moisture content, which was 53% of the pre-stabilized CR OMC. The samples were not wrapped or otherwise isolated from the surrounding CR moisture. The temperature range of the simulated in-pavement curing was 10–30 °C, selected to reflect typical post-construction base layer temperatures. It is acknowledged that the simulated in-pavement curing pro-tocol was not necessarily a true reflection of in-pavement conditions, but it is expected to be substantially more representative than the accelerated laboratory oven curing.

The design aircraft used for FBS pavement design was a B737-800 at its standard tyre pressure (1.41 MPa) and maximum take-off mass (78 t). A total of 100,000 passes were assumed over a 20-year design life, intended to reflect a typical major regional airport in Australia. Lateral aircraft wander was ignored to simplify the modelling.

The pavement structure was selected to be representative of a regional airport in Australia to be upgraded by FBS to accommodate the B737-800 aircraft on a regular basis. The modelled pavement included an asphalt surface, FBB base and NG sub-base over the subgrade (Table 2) and all materials were characterized according to Australian practice (AAA 2017). The FBB modulus was varied to reflect the material testing results and the required thickness of the NG sub-base was adjusted to achieve the required pavement strength, again following normal Australian airport pavement design practices.

Table 2. Pavement composition and materials

Layer/course	Material	Thickness (mm)	Modulus (MPa)	Poisson's ratio
Surface	Airport asphalt	50	1500	0.40
Base	FBB	300	Varied	0.30
Sub-base	NG	As required	Automatic	Automatic
Subgrade	Natural CBR 4%	Infinite	40	0.40

The FBB base modulus was determined by laboratory testing. The NG sub-base thickness was calculated to provide a pavement structured estimated to fail at the end of the modelled aircraft traffic. NG was sub-layered and modulus/Poisson's ratio values assigned according to the Barker and Brabston (1975) method.

3.2 Results

The duplicate/triplicate cured and saturated modulus values for the insitu, exsitu and laboratory produced FBB are shown in Fig. 4. The average exsitu produced and laboratory cured FBB modulus results are in Fig. 5, along with the exsitu produced simulated in-pavement cured modulus results.

4 Discussion

4.1 Effect of Production Method on Modulus

The pre-cured modulus values were not significantly different for the three production processes (p-values 0.83 and 0.30). This reflects the high influence of the inter-particle friction provided by the aggregate on the uncured properties of FBB. However, with curing, the differences between the modulus associated with each production method was significant (p-values < 0.01) (Fig. 6). On average, the exsitu produced modulus values were 58% (cured) and 55% (saturated) of the laboratory produced FBB, while the insitu produced modulus values were 43% (cured) and 35% (saturated) of the laboratory produced values. Furthermore, when the cured and saturated results were combined, the

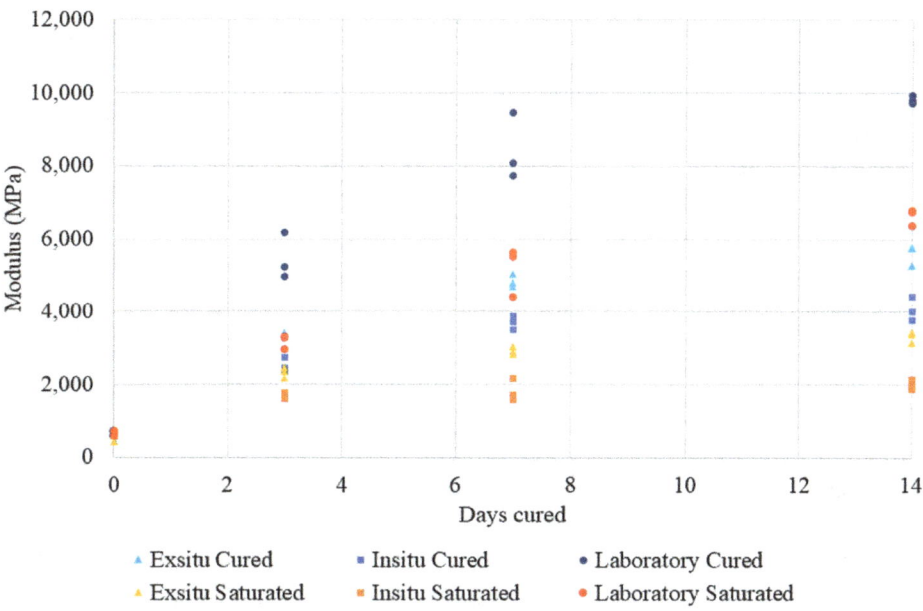

Fig. 4. FBB modulus results for different production methods

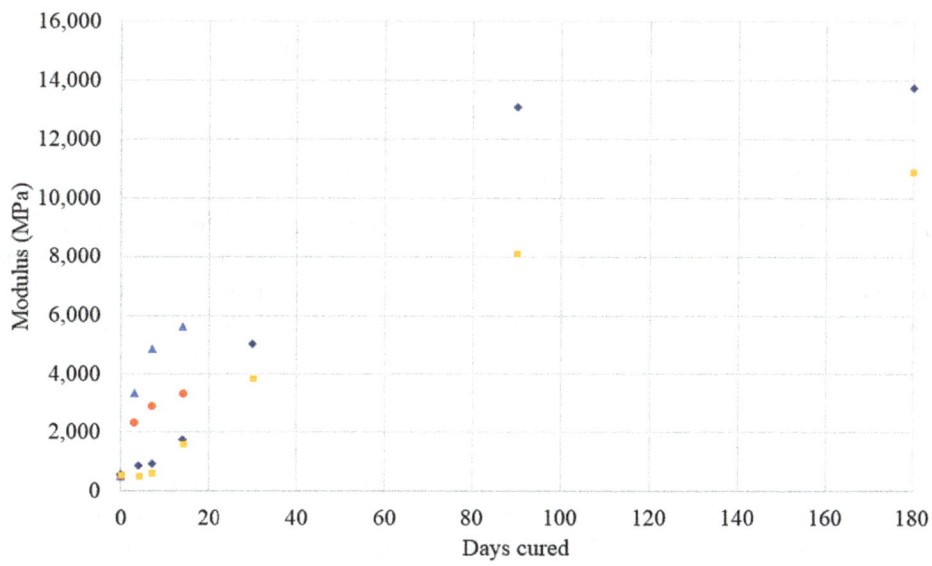

Fig. 5. FBB modulus results for laboratory and simulated in-pavement curing

exsitu stabilized modulus values were, on average, 39% higher than the equivalent insitu produced material modulus values (Fig. 7). This is expected to reflect the mixing energy and distribution of the different production processes, which produced visually different materials (Fig. 2), despite the nominally identical mixture design.

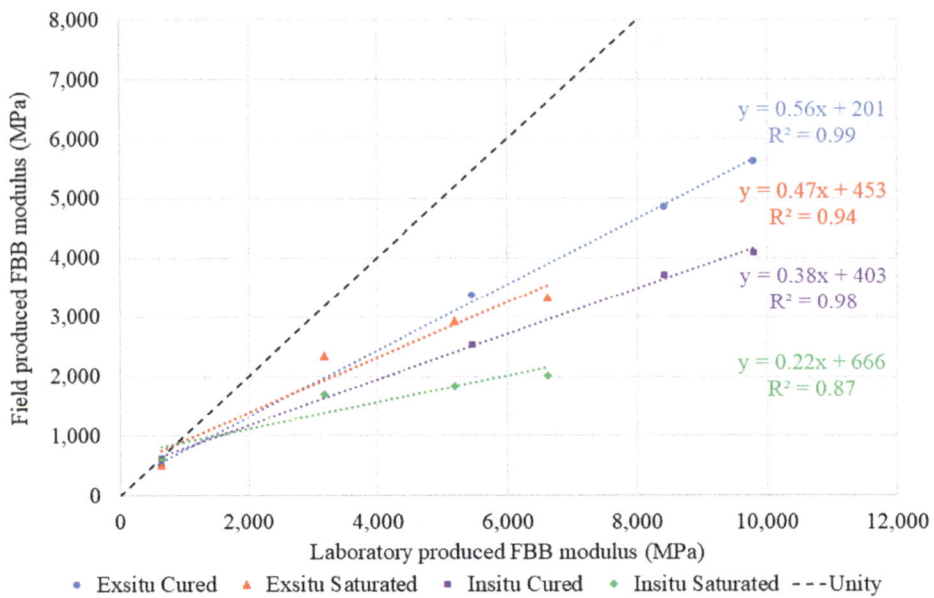

Fig. 6. Field produced FBB modulus compared to laboratory produced modulus

4.2 Effect of Curing Method on Modulus

The increase in average FBB modulus with curing time is clearly shown in Fig. 8. The 14 day modulus results were 3–15 times the pre-cured results for the various production methods. The majority of the modulus increase occurred within the first three days and thereafter the increase in modulus slowed. For example, the increase in cured modulus from day 3 to day 7 was 44–54% while the increase from day 7 to day 14 was just 10–16%.

The difference in modulus associated with accelerated laboratory curing and the simulated in-pavement curing is also clear, as shown in Fig. 9. As stated above, the modulus increased rapidly during the first three days in the 40 °C oven and the rate of increase slowed. This reflects the rapid drying out of the samples in the oven. In contrast, the samples cured in the simulated pavement gained modulus much more slowly due to the slower drying out associated with the lower temperature and the moisture retained in the surrounding aggregate.

The three day accelerated laboratory cured modulus value was exceeded by the simulated in-pavement curing modulus in 15–25 days of simulated curing. Furthermore, the 14 day accelerated cured modulus was exceeded by the simulated curing modulus n 25–35 days of simulated curing. The comparison of the two curing conditions indicates that FBS pavements are susceptible to overloading for the first two to three weeks

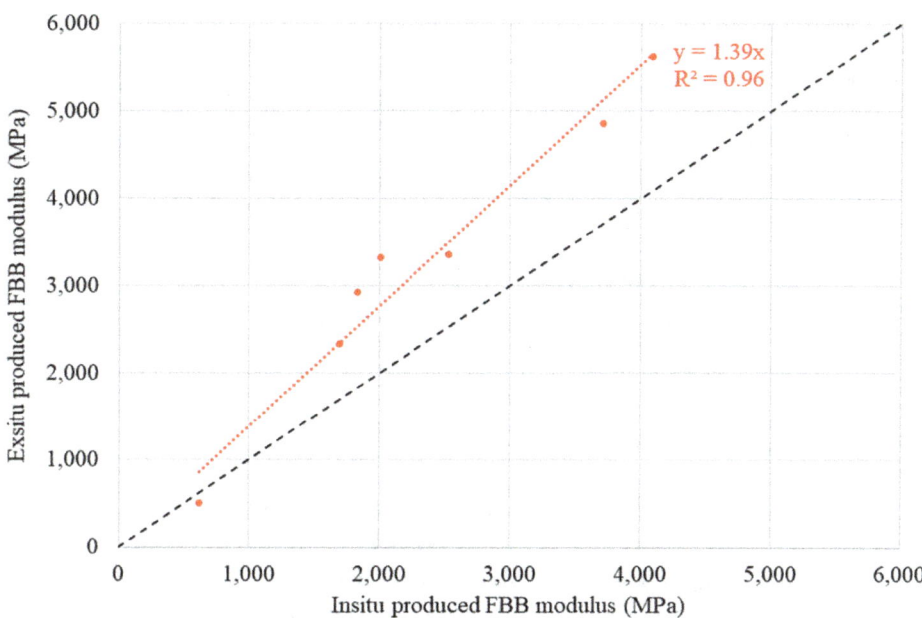

Fig. 7. Insitu produced FBB modulus compared to Exsitu produced FBB modulus

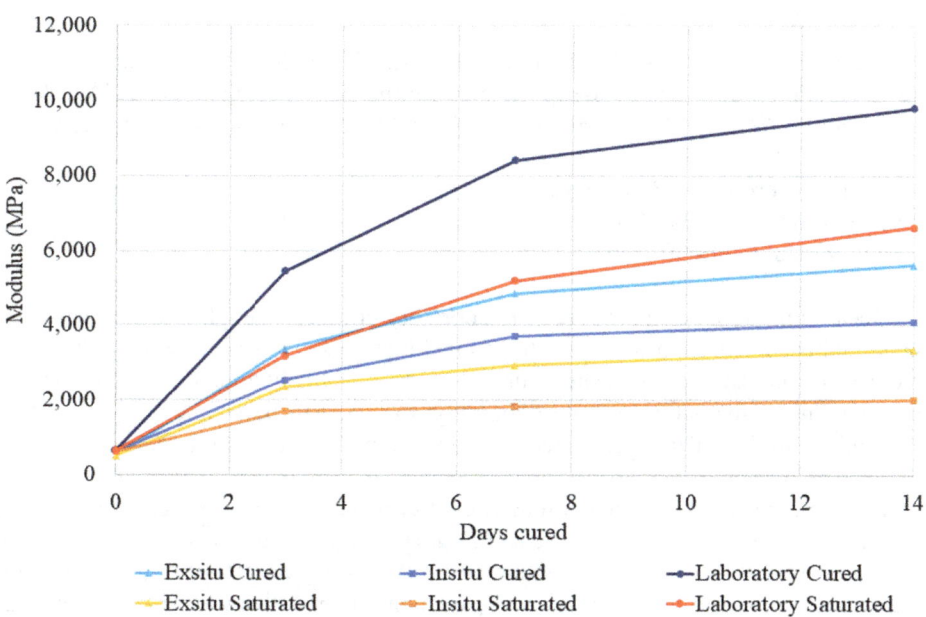

Fig. 8. Laboratory cured FBB modulus with curing time for different FBB production methods

after construction, after which they are expected to significantly exceed their intended strength.

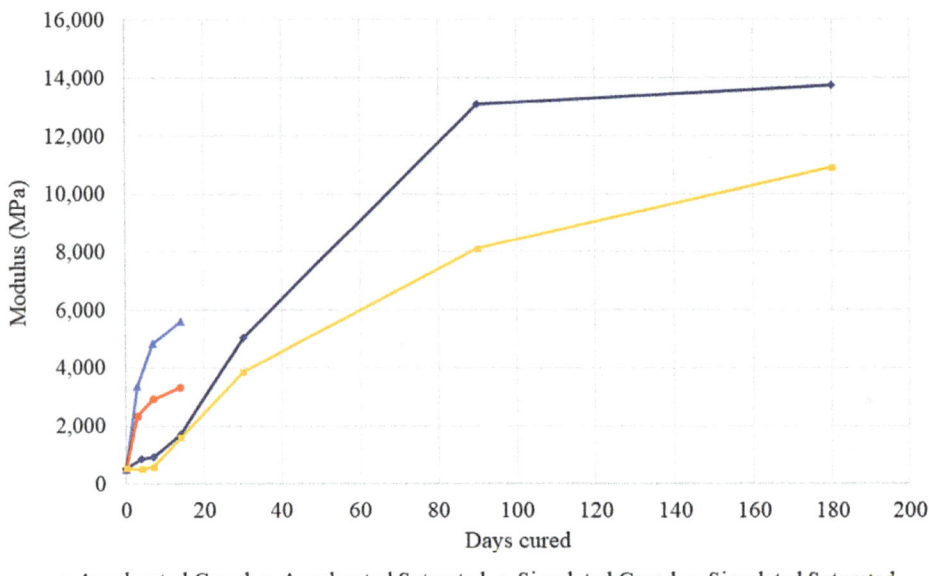

—◆—Accelerated Cured —◆—Accelerated Saturated —◆—Simulated Cured —◆—Simulated Saturated

Fig. 9. Exsitu produced FBB modulus for accelerated and simulating curing time

4.3 Characteristic Modulus Values

The Australian guidance on FBS requires the saturated modulus after three days of accelerated laboratory curing to be used for pavement thickness design purposes (Austroads 2019). For the FBB produced using the three production processes, those values were:

- Laboratory produced. 3,180 MPa.
- Exsitu produced. 2,340 MPa.
- Insitu produced. 1,690 MPa.

Because mixture design is based on laboratory produced samples, it over-estimates the modulus compared to actual exsitu and insitu produced materials cured in the same manner. In contrast, the use of the saturated modulus is likely to underestimate the true modulus due to the actual moisture condition expected to be significantly less harsh than full saturation. This is expected to offset the over-estimation associated with the accelerated curing.

Furthermore, the saturated modulus after three days of accelerated curing is intended to represent medium-term in-pavement curing. However, as stated above, it is more representative of 15–25 in the pavement and even the 14 day accelerated curing modulus was representative of only 25–35 days in-pavement. After 180 days of in-pavement curing, which is just 1.25% of a 20 year design life, the saturated modulus was 10,910 MPa. This is more realistic of medium-term and long-term performance. However, if the pavement is to be trafficked the day after construction, as is common for expedient aircraft pavement upgrade works, then assuming a modulus based on 180 days of in-pavement curing would be unreasonable.

Table 3 summarizes the various modulus values that might be considered to be appropriate for use in pavement thickness design, including the laboratory produced saturated values after three days of accelerated curing, which is prescribed by the Australian guidance (Austroads 2019). The initial modulus is appropriate for early trafficking, but it is unlikely that the FBB would be saturated immediately after construction. Consequently, the initial modulus was tested at the initial moisture condition and the same value was used to represent both the initial cured and saturated conditions. In contrast, it would be appropriate to base the long-term modulus on the 180 day in-pavement curing value, either unsaturated (in an arid region) or saturated (in a tropical region).

Table 3. Potential modulus values (MPa) for pavement design

Modulus basis	Exsitu	Insitu	Laboratory
Before curing	510	630	650
Saturated after 3 days of accelerated curing	2,340	1,690	3,180
Unsaturated after 3 days of accelerated curing	3,370	2,530	5,460
Saturated after 14 days of accelerated curing	3,330	2,010	6,630
Unsaturated after 14 days of accelerated curing	5,620	4,090	9,810
Saturated after 90 days of simulated curing	8,110	–	–
Unsaturated after 90 days of simulated curing	13,100	–	–
Saturated after 180 days of simulated curing	10,910	–	–
Unsaturated after 180 days of simulated curing	13,760	–	–

4.4   Effect of Modulus on Pavement Thickness and Life

APSDS was used to determine the pavement thickness based on the laboratory produced and three day cured and saturated modulus after three days curing (3,180 MPa). This was referred to as the reference thickness. Alternate FBB modulus values were selected, intended to be representative of the values in Table 3, and the associated sub-base thicknesses were also determined for each FBB modulus values (Table 4). The reference pavement thickness was then retained and the change in predicted pavement life was determined for each alternate FBB modulus value (Table 4). In all cases, the pavement life was based only on subgrade rutting as the failure mode. The potential for fatigue of the FBB and/or the asphalt surface was omitted, as is commonly the case for airport pavement thickness design (AAA 2017).

The total pavement thickness reduced, approximately linearly, from 1,142 mm (for 600 MPa FBB modulus) to just 350 mm (for 7,000 MPa). The 350 mm total pavement thickness was equal to the surface asphalt and base course thicknesses and effectively identified that no sub-base thickness was required for 7,000 MPa (or greater) FBB modulus, as shown in Fig. 10. Furthermore, for the reference thickness, the effect of FBB modulus on relative pavement life was significant, with the pavement predicted to

Table 4. Pavement thickness and relative life for FBB modulus values

FBB modulus (MPa)	Sub-base thickness required (mm)	CDF at reference thickness
600	792	33.1
1,500	705	6.7
2,000	665	3.5
2,500	628	2.0
3,000	596	1.2
3,180	582	1.0
4,000	515	0.47
5,000	420	0.21
6,000	247	0.10
7,000	<50	0.06
9,000	<50	0.02
11,000	<50	0.01
14,000	<50	<0.01

3,180 MPa is italicized as the reference modulus, with its associated reference thickness and CDF of 1.0.

last just 3% of the design life when the FBB modulus was 600 MPa, but 1,100 times the design life if the modulus was 11,000 MPa (Fig. 11). It follows that the characteristic FBB modulus is critical to pavement thickness design and the majority of pavement damage is expected to occur early in the pavement's life, when the FBB modulus is low, with further pavement damage being insignificant once the characteristic modulus value is exceeded.

As shown in Fig. 8, the three production methods produced significantly different modulus values. The differences were relatively similar for cured and saturated conditions and for various curing times. However, the saturated modulus after three days of accelerated curing is the standard for Australia FBB characterization and pavement thickness design. The modulus values and associated reference pavement thickness CDF values are summarized in Table 5. Using the laboratory mixer to produce FBB for pavement thickness design characterization significantly overestimated the FBB modulus, resulting in an under-strength pavement structure. As a result, the reference thickness pavement was predicted to fail in just 8.5 years (exsitu produced) or 3.9 years (insitu produced) into the 20-year design life. However, this analysis was based on standard accelerated laboratory curing and omitted the effect of FBB age on FBB modulus and the subsequent impact on pavement damage.

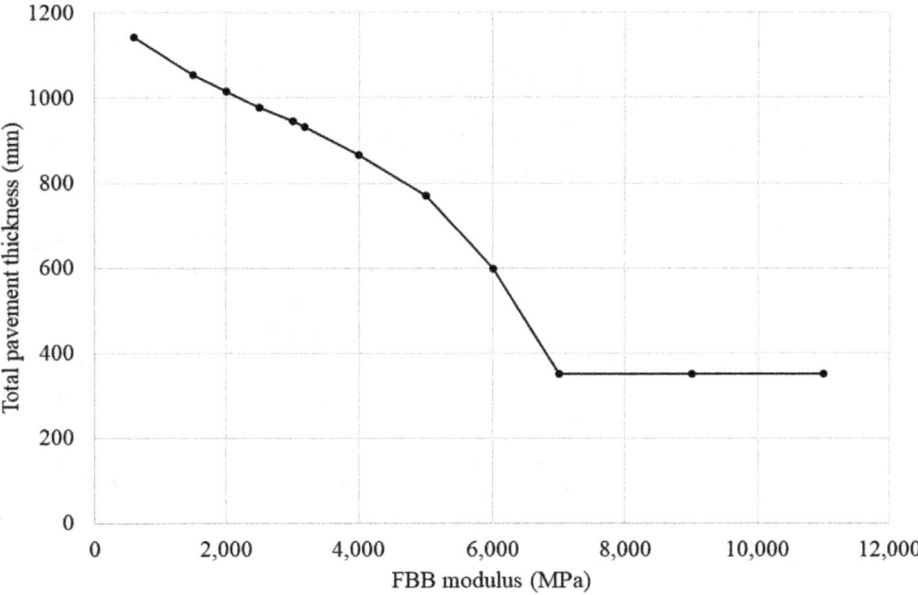

Fig. 10. Effect of FBB modulus on required pavement thickness

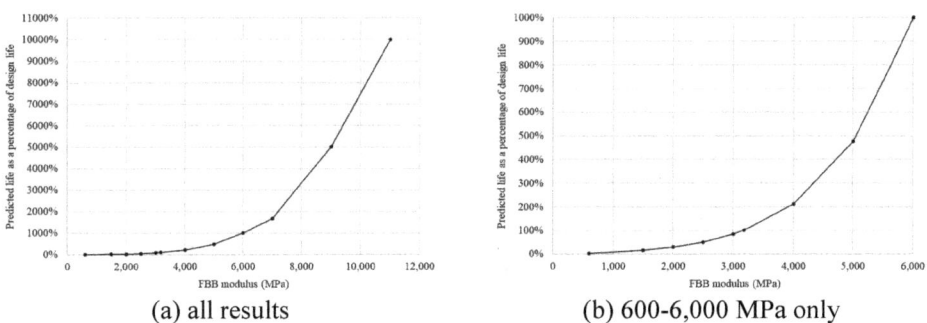

(a) all results   (b) 600-6,000 MPa only

Fig. 11. Effect of FBB modulus on reference pavement relative life

Table 5. Production method FBB modulus and pavement CDF values

FBB production method	3 day accelerated cured and saturated modulus (MPa)	CDF at reference pavement thickness
Laboratory mixed	3,180	1.00
Exsitu pugmil	2,340	2.36
Insitu stabilizer	1,690	5.16

4.5 Pavement Damage with Age

As discussed above, layered elastic pavement design generally assumes a constant modulus over the design life of the pavement. This means that the incremental accumulation of damage is modelled to occur at a constant rate with regard to the cumulative number of aircraft loadings. However, if the modulus changes significantly over the pavement life, then the rate of damage cumulation will change as well.

To determine the rate of damage cumulation two predictive models of FBB modulus gain over time were developed from the exsitu produced, simulated in-pavement curing results (Fig. 4). One model predicted the cured modulus and the other model considered the saturated modulus. The adopted models included a constant modulus for the first five days, then a linearly increasing modulus up to day 89, followed by a constant modulus from day 90 and thereafter. The initial (days 0–5) constant modulus values were based on the average modulus measured during this period, while the latter (days 90 and thereafter) constant modulus values were based on the average of the values measured at day 90 and day 180. The increasing modulus from day 6 to day 89 were each based on a linear fit between the day 5 and day 90 modulus values, as shown in Fig. 12 and detailed as Eq. 1 (cured) and Eq. 2 (saturated).

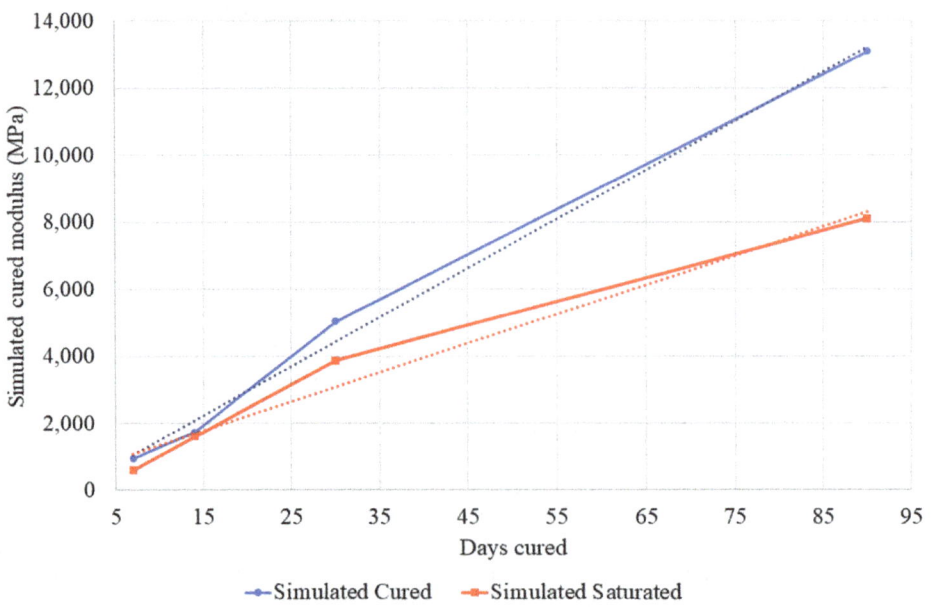

Fig. 12. Linear models for modulus with age from day 6 to day 89

$$E(cured) = \begin{cases} 781 & (days\,0-5) \\ 150.6 \times d - 122 & (days\,6-89) \\ 13,429 & (days\,90-7,200) \end{cases} \tag{1}$$

$$E(saturated) = \begin{cases} 557 & (days\ 0-5) \\ 106.6 \times d - 82 & (days\ 6-89) \\ 9508 & (days\ 90-7,200) \end{cases} \tag{2}$$

Where $E(cured)$ = modulus for cured FBB
   $E(saturated)$ = modulus for saturated FBB
   d = days of curing

A predictive model was then developed for pavement damage as a function of FBB modulus. Equation 3 was developed as an approximation of the relationship shown in Fig. 11, with an R^2 value of 0.99, indicating good agreement with the actual values. The material-specific modulus gain models (Eq. 1 and Eq. 2) were then combined with the material-specific damage model (Eq. 3) to predict pavement damage over the life of the pavement, assuming a constant rate of trafficking, but an increasing material-specific FBB modulus. Separate damage predictions were developed for the cured and saturated modulus values over time.

$$CD = \left(1.29 \times 10^{-10} \times E^3 - 7.06 \times 10^{-7} \times E^2 + 1.33 \times 10^{-3} \times E - 0.49\right)^{-1}$$

$$\tag{3}$$

Where CD = Cumulative damage
   E = FBB modulus (cured or saturated)

The cumulative damage was finally adjusted to reflect the portion of traffic occurring each day, which was 100,000 passes over 20 years, equal to 13.7 passes per day, and then summed over the 7,300 days in the 20-year design life, to produce a cumulative damage factor, equivalent to the CDF values in Table 4. The daily damage was high during the initial (constant) modulus phase and then reduced to an insignificant value by day 90, after which the additional damage was negligible (Fig. 13). When the daily damage was summed over the design life of the pavement, the cured modulus CDF was 1.14 and the saturated modulus CDF was 2.69 (Fig. 14). This indicates that when the simulated in-pavement FBB modulus gain with curing time was realistically modelled in pavement design, the cured modulus values resulted in a comparable CDF value to that implied by the laboratory produced FBB saturated modulus value after three days of accelerated laboratory curing, which is typically used for FBB characterization. That is, a pavement thickness determined using the modulus from standard characterization protocols, is a reasonable proxy for the predicted life of a pavement based on the modulus associated with exsitu FBB production and realistic in-pavement curing conditions.

The above analysis was based on uniform aircraft traffic loadings over the pavement life. However, it is clear that the majority of the pavement damage occurred early in the pavement life when the FBB modulus was low. Consequently, if the aircraft traffic changed over time, the outcome of the analysis would be significantly different. For example, if the pavement was not trafficked until the FBB was 90 days old, the pavement would never be significantly damaged because the FBB modulus was predicted to be

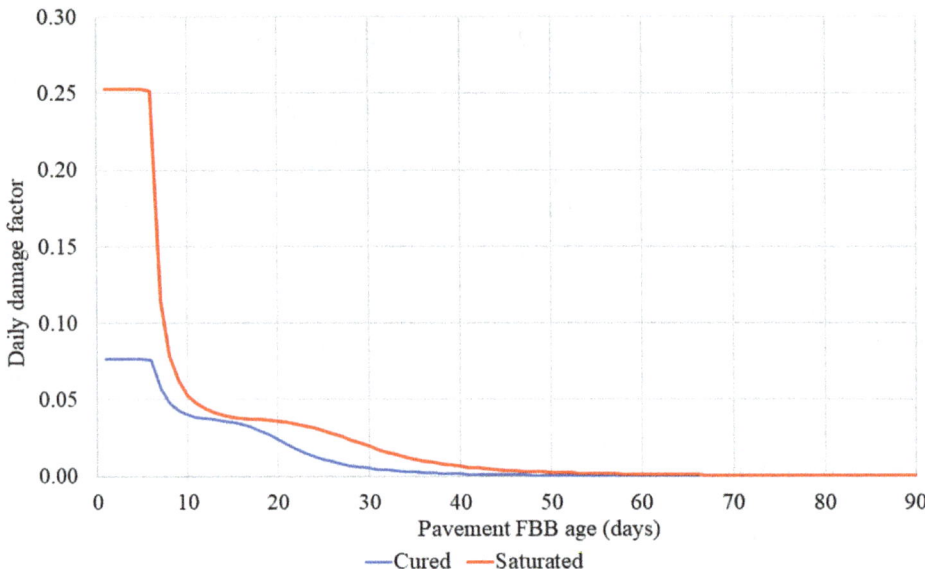

Fig. 13. Daily pavement damage based on estimated FBB modulus gain

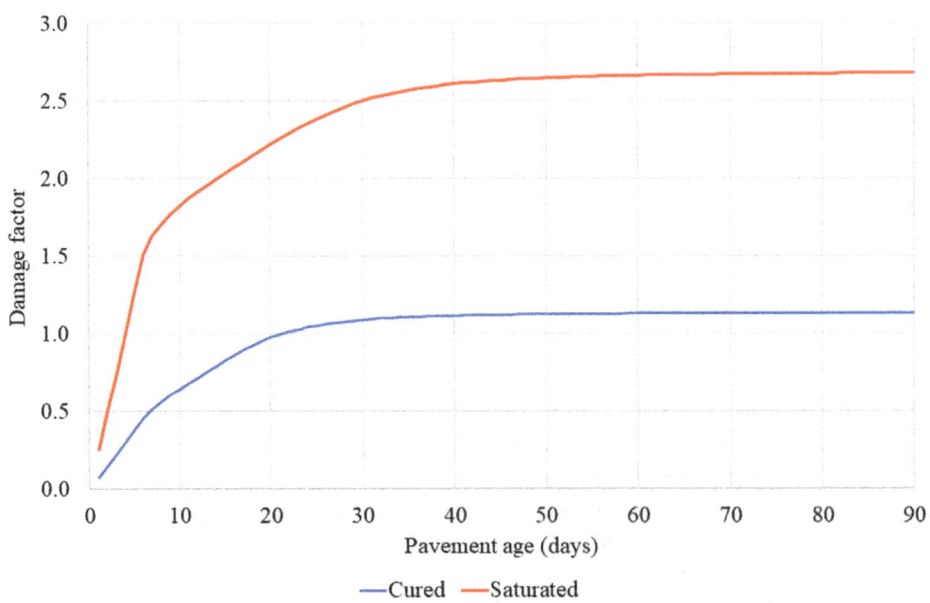

Fig. 14. Cumulative pavement damage based on estimated FBB modulus gain

above 8,000 MPa at that time. Conversely, a designer would be justified in selecting a higher FBB modulus for layered elastic thickness determination if it was known that the FBB would be allowed to cure for 90 days before trafficking.

5 Conclusions

The modulus of the field produced FBB increased significantly during the first 90 days after production and the laboratory production and curing protocols used in Australia were not representative of field production and simulated in-pavement curing conditions. This resulted in the laboratory determined characteristic FBB modulus values significantly over-estimating the actual in-pavement modulus for the first 5–10 days following FBB production, but significantly under-estimating the actual in-pavement modulus achieved after 90 days of curing. Laboratory characterization protocols for FBB should be modified to be more representative of field production and curing conditions in the future. Based on layered elastic modelling, it was concluded that pavement thickness determined by laboratory production and curing protocols was a reasonable proxy for the predicted life of a pavement comprising exsitu produced FBB and subject to realistic curing conditions in an unsaturated environment. It was also found that more than 80% of the predicted pavement damage occurred in the first 20 days after FBB production and only negligible damage occurred after the FBB was 90 days old. It is therefore recommended that FBB remains untrafficked for 7–12 days after production, wherever possible, and that higher FBB modulus values and thinner pavements are likely to perform adequately in situations where the FBB is protected from traffic loading for more than 14 days following construction.

References

AAA: Airfield Pavement Essential, Airport Practice Note 12. Australian Airports Association. Canberra, Australian Capital Territory, Australia, April 2017

Austroads: Resilient Modulus of Foamed Bitumen Stabilised Materials. Test Method AGPT/T305. Austroads. Sydney, New South Wales, Australia, October 2017

Austroads: Design and Performance of Foamed Bitumen Stabilised Pavements. Technical Report AP-T336-18, Austroads. Sydney, New South Wales, Australia, July 2018

Austroads: Guide to pavement technology: part 4D: stabilised materials. Austroads publication AGPT04D-19. Austroads. Sydney, New South Wales, Australia, April 2019

Barker, W.R., Brabston, W.N.: Development of a Structural Design Procedure for Flexible Aircraft Pavements. Report No S77-17. US Army Corps of Engineers. Waterways Experiment Station, Vicksburg, Mississippi, USA (1975)

Fu, P., Jones, D., Harvey, J.T.: Micromechanics of the effects of mixing moisture on foamed asphalt mix properties. J. Mater. Civ. Eng. 22(10) (2010)

Huang, Y.H.: Pavement Analysis and Design. Prentice-Hall, New Jersey (1993)

Morgan, J.: Prediction of soil moisture conditions for pavement design. In: 7th Biennial AARB Conference. Australian Road Research Board, Melbourne, Victoria, Australia, July 1972

Muller, W.: Characterising moisture within unbound granular pavements using multi-offset ground penetrating radar. A thesis submitted for the award of Doctor of Philosophy. School of Civil Engineering, University of Queensland, St Lucia, Queensland, Australia (2017)

Soufi-Sabbagh, J.: The utilisation of EME2 asphalt at Brisbane international Airport. Pavement and Lighting Forum. Brisbane, Queensland, Australia, 12-14 May 2018

TMR: Unbound Pavements. Technical Specification MRTS09. Department of Transport and Main Roads, Government of Queensland. Brisbane, Queensland, Australia, November 2019

TMR: Materials Testing Manual. Edition 5, Amendment 3. Department of Transport and Main Roads, January 2020

Wardle, L.J.: Program CIRCLY User's Manual. Commonwealth Scientific and Industrial Research Organisation, Australia, Division of Applied, Geomechanics, Geomechanics Computer Program No 2 (1977)

Wardle, L.J., Rodway, B.: Advanced design of flexible aircraft pavements. In: 24th ARRB Conference, Melbourne, Victoria, Australia, 12–15 October 2010

Weir, T.: Experimental investigation of the influence of production processes and curing on foamed bitumen stabilised base course materials. A thesis submitted for the award of the degree Master of Science (Engineering). University of the Sunshine Coast, Sippy Downs, Queensland, Australia, March 2020

White, G.: A sensitivity analysis of APSDS, an Australian mechanistic design tool for flexible aircraft pavement thickness determination. In: First European Aircraft Pavement Workshop. Amsterdam, Netherlands, 11–12 May 2005

White, G.: A probabilistic approach to flexible aircraft pavement thickness determination. In: 8th International Bearing Capacity of Road, Railways and Airfields Conference. Champaign, Illinois, USA, 29 June–2 July 2009

White, G.: Foamed bitumen stabilisation for Australian airports. In: Airfield Engineering and Maintenance Summit. Furama Riverfront, Singapore, 25–28 March 2014

White, G.: Expedient runway upgrade technologies. In: 10th International Conference on the Bearing Capacity of Roads, Railways and Airfields, Athens, Greece, 28–30 June 2017

White, G.: Foamed bitumen base for airport pavements. In: 28th ARRB International Conference. Brisbane, Queensland, Australia, 30 April–2 May 2018

White, G., Fairweather, H., Jamshidi, A.: Sustainable runway pavement rehabilitation: a case study of an Australian airport. J. Clean. Prod. **204**, 380–389 (2018)

White, G., Kelly, G., Fairweather, H., Jamshidi, A.: Theoretical socio-enviro-financial cost analysis of equivalent flexible aircraft pavement structures. In: 99th Annual Meeting of the Transportation Research Board. Washington, District of Columbia, USA, 12–16 January 2020

Wirtgen: Wirtgen Cold Recycling Technology. Wirtgen, Windhagen (2012)

Using Recycled Concrete as a Replacement for Coarse Aggregate in the Production of Green Concrete

Mohamed Tarek El-Hawary[1,2(✉)], Carsten Koenke[2], Amr El-Nemr[3], and Nagy F. Hanna[3]

[1] German University in Cairo, Cairo, Egypt
elhawary.mth@gmail.com
[2] Bauhaus-Universität Weimar and MFPA Weimar, Weimar, Germany
Carsten.koenke@uni-weimar.de
[3] Civil Engineering Department, German University in Cairo, Cairo, Egypt
amr.elnemr@guc.edu.eg, nagy.hanna@gu.edu.eg

Abstract. Reducing the environmental impact and enhancing the properties of concrete is of great significance environmentally and economically. Using sustainable construction materials coming from construction wastes like recycled concrete that comes out from demolishing or renovating of existing structures could help in producing sustainable materials of good quality and low cost. The use of recycled concrete improved the concrete quality and enhanced the mechanical and microstructural properties of concrete. Tests and research done on this type of sustainable concrete proved the applicability of usage of recycled concrete aggregates as a full replacement of coarse aggregate by using different grain sizes of natural and recycled coarse aggregates. The desired concrete strength as per the concrete mix design is 35 MPa with fixed water to cement ratio of 0.48. The concrete mechanical properties increased clearly after replacing the natural coarse aggregates by recycled concrete aggregates. This research studied the concrete density, workability, temperature, compressive strength, split tensile strength, flexural strength, and microstructural analysis; which is the main focus of this paper.

Keywords: Green concrete · Recycled concrete aggregates · Sustainable construction materials · Mechanical properties · Microstructural analysis · Fresh concrete properties · Hardened concrete properties

1 Introduction

Concrete production and industry is the main concern in research nowadays, taking into account the sustainability of the environment and the economic benefits from such researches. From the environmental aspect, the ceramic wastes contribute with huge percentage -about 30% from the production of ceramic tiles comes as waste, (Raval

---

N. F. Hanna—Former Professor at Mataria University.

© The Author(s), under exclusive license to Springer Nature Switzerland AG 2021
H. Shehata and S. El-Badawy (Eds.): *Sustainable Issues in Infrastructure Engineering*, SUCI, pp. 127–151, 2021.
https://doi.org/10.1007/978-3-030-62586-3_9

et al. 2013) - in the construction waste, and also the ceramic industry waste. The use of these wastes enables us to get rid of these wastes in a useful way, instead of the traditional ways that are harmful to the environment. From the economic aspect, using ceramic as replacement to the aggregate has its own benefits as it reduces the cost of concrete production and also it is a lightweight material, so it decreases the total weight of the structure. In addition to lowering the cost of concrete mix itself, this will enable the designers to decrease the foundation cost of the buildings as a result of using lighter weight concrete. Using concrete waste could save energy, reduce the manufacturing cost of concrete, and also save the environment, (Medina et al. 2012).

The replacement of aggregate in concrete can be done using several materials such as ceramic which is the used material in this research, bricks (Adams 2012), oil palms shells (Basri et al. 1999), tire rubber (Nehdi and Khan 2001), quarry wastes (Nataraja et al. 2001), building rubbles (Khalaf and DeVenny 2004), vitrified soil aggregate (Palmquist et al. 2001), sand stone (Kumar et al. 2007), basaltic pumice (Binici 2007), and electric arc furnace slag (Manso 2006). The use of all these materials has economical, and environmental benefits by reducing the cost of concrete manufacture and decreasing the landfill demand for such solid waste materials.

The use of construction wastes could be categorized into three main groups, coarse aggregate replacement for concrete mixes, fine aggregate replacement for cement mortar, and cement replacement for cement mortar. The used wastes are crushed using grinder. Usage of construction wastes in construction industry is of significant importance from the environmental and economic aspects; this requires implementing tests and experiments on the different types of wastes after classifying it, to be examined for introducing this type of material to the new sustainable construction materials technologies. By classifying the construction wastes the following wastes were found to be of different characteristics regarding its chemical composition, shapes and forms, strength, and many other aspects. After studying the market and the construction industry especially in Egypt, the first material that was tested in the previous research was the ceramic waste, and its different properties as a specific material and also as new sustainable construction material. The scope of this research is to continue testing the construction wastes and to validate different types of materials and relate their behaviour to each other, reaching a design methodology or a design technique for the green construction materials.

Marble waste is one of the strongest materials that we can predict a good behaviour regarding the concrete strength; if it is introduced to the concrete design. Marble manufacturing and industry produces a huge amount of marble waste coming from two main resources. The first and the main resource is coming from the preparation of the marble and converting the raw blocks into the traditional tiles or sheets that is used in floorings, finishing, or any other application, which produces about 70% of the mineral as a waste during the mining, processing and polishing stages (Hebhoub et al. 2011). While producing marble the most important aspect is the aesthetics and hardness according to the type of application or how it would be used. It happens a lot that if the block is not looking good or aesthetically not qualified for the market demands and specifications that the whole block would not be used and it is considered as a waste. The other source comes from demolishing buildings and old or used tiles or sheets of marble all of these quantities could be considered as a waste and should be used in a way or another. The main idea is to use the marble waste as replacement for the natural aggregates in the concrete manufacture and replacing the cement by the marble dust which is less than 1 mm size.

2   Problem Statement and Motivation

Hebhoub et al. (2011), using the marble waste in the concrete manufacturing solves this problem environmentally and economically, especially with a 75% replacement value -of any formulation- under constant water to cement ratio. The use of the marble waste as replacement for coarse and fine aggregate separately or together leads to an increase in the concrete mechanical properties. Not only enhancing the concrete properties, the use of ceramic tiles waste also reduces the cost of concrete production and saves the environment from the huge amount of wastes that couldn't be easily get rid of, and save the natural non-renewable materials and resources as well.

3   Background and Previous Research

Researches were done to examine the sustainable concrete properties using different types of wastes. The characteristics of the produced concrete were handled in different ways; some researches studied only the mechanical properties or studied one of the mechanical properties only neglecting the other properties. After collecting different results and interpretations from previous researches the following data were found.

El-Hawary et al. (2019), introduced sustainable concrete using marble aggregates with different sizes as a replacement for the natural aggregates in concrete. The concrete mix was done according to the ACI standards and with water to cement ratio of 0.48. replacement percentages of natural aggregates with marble aggregates varied from 0% till reaching 100%. The research proved the applicability of using the full replacement of natural aggregates by marble aggregates; which improved the concrete mechanical and microstructural properties.

Binici et al. (2008), investigated the cement and aggregate replacement with marble waste in concrete with constant water to cement ratios. The results showed improvements in the mechanical and chemical properties of cement and concrete after adding the marble waste to the mix with fixed water to cement ratio of 0.4, shown in Table 1.

Ergun (2011), added marble to the concrete as a replacement for the coarse aggregate; which enhanced the concrete mechanical properties, workability and chemical resistance if compared to the traditional concrete as shown in Table 2. The reason behind using the marble in producing more durable concrete is not only economic but also environmental; for example in Turkey, which has huge amounts of marble waste due to the availability of the raw material. Experiments and tests done on replacing natural aggregates with recycled marble sand and recycled marble gravel showed better concrete mechanical properties and a significant increase in the concrete compressive and tensile strength, especially for the 25%, 50%, and 75% of replacement of natural aggregates by marble waste aggregates. The workability could be improved using the optimized water to cement ratio as mentioned in the results.

Table 1. Concrete mechanical properties after different aging Binici et al.

Mixture no.	Density hardened concrete	Compressive strength MPa					Flexural strength MPa	Split tensile strength MPa	7 vs 28 days	Average	28 vs 90 days	Average
		1 day	7 days	28 days	90 days	356 days	28 days	28 days				
MC1-1	2353	29.1	38.4	44.4	50	57.4	6.6	3.4	13.51	13.84	11.20	9.89
MC1-2	2355	30	37.9	44.6	49	57.1	6.2	3.5	15.02		8.98	
MC1-3	2357	28.5	38.2	43.9	48.5	28.2	6.4	3.1	12.98		9.48	
MC1 average	2355	29.2	38.2	44.3	49.2	57.6	6.4	3.3	13.77		9.96	
MC2-1	2333	31	41.1	46.6	55	62.6	7.2	3.6	11.80	14.94	15.27	12.73
MC2-2	2344	31.1	38.3	47.8	54.2	61.9	6.9	3.3	19.87		11.81	
MC2-3	2340	29.6	41	47.2	53.1	61.8	6.3	3.4	13.14		11.11	
MC2 average	2339	30.1	40.1	47.2	54.1	62.1	6.9	3.5	15.04		12.75	
GC1-1	2352	29.2	36.8	44.5	50.1	55.9	6.5	3.3	17.30	13.77	11.18	11.89
GC1-2	2344	27.8	35.9	42.8	49.4	56.8	6.1	3.1	16.12		13.36	
GC1-3	2332	28.1	39.7	43.1	48.5	57.7	6.3	3.2	7.89		11.13	
GC1 average	2343	28.4	37.5	43.5	49.3	56.8	6.3	3.2	13.79		11.76	
GC2-1	2332	30.7	38.3	42.1	49.5	57.8	6.6	3.5	9.03	12.42	14.95	12.77

(continued)

Table 1. (*continued*)

Mixture no.	Density hardened concrete	Compressive strength MPa					Flexural strength MPa	Split tensile strength MPa	7 vs 28 days	Average	28 vs 90 days	Average
		1 day	7 days	28 days	90 days	356 days	28 days	28 days				
GC2-2	2328	28.7	40.1	46	51.4	59.6	6.2	3.3	12.83		10.51	
GC2-3	2315	29.4	37.3	44.1	50.6	60.7	6.5	3.3	15.42		12.85	
GC2 average	2325	29.6	38.6	44	50.5	59.4	6.5	3.4	12.27		12.87	
C1-1	2402	8.5	16.1	26.2	31.5	37.9	4.1	2.2	38.55	34.68	16.83	22.18
C1-2	2381	8	16	24.1	33.2	37.5	3.8	2.1	33.61		27.41	
C1-3	2390	8.6	17.1	25.1	32.3	34.9	3.9	1.9	31.87		22.29	
C1 average	2391	8.4	16.4	25.1	32.3	36.8	3.9	2.1	34.66		22.29	
C2-1	2388	14.1	29.3	33.9	42	48.4	4.4	2.4	13.57	19.38	19.29	14.15
C2-2	2373	13.5	27.8	36.1	43.3	46.3	4.2	2.3	22.99		16.63	
C2-3	2369	13.6	28	35.7	38.2	46.9	4.2	2.5	21.57		6.54	
C2 average	2377	13.7	28.4	35.2	41.2	47.2	4.3	2.4	19.32		14.56	

Table 2. Concrete mechanical properties Ergun (2011)

	Compressive strength MPa			7 vs 28 days	28 vs 90 days	Flexural strength MPa			7 vs 28 days	28 vs 90 days
	7 days	28 days	90 days			7 days	28 days	90 days		
Control	27.2	35.4	35.6	23.16	0.56	5.1	5.3	5.7	3.8	7.0
D5	29.8	36.1	37.9	17.45	4.75	5	5.1	5.9	2.0	13.6
D7.5	33.4	40.4	41.5	17.33	2.65	5.1	5.2	5.5	1.9	5.5
D10	34.7	42.6	46.1	18.54	7.59	5.1	5.3	5.8	3.8	8.6
M5	30.3	39.4	40.9	23.10	3.67	5.1	5.3	6	3.8	11.7
M7.5	29.5	39.9	41.1	26.07	2.92	5.1	5.1	5.6	0.0	8.9
M10	25.8	31.1	31.3	17.04	0.64	4.8	5	5.1	4.0	2.0
D5M5	34.2	39.7	42.4	13.85	6.37	5.4	5.5	6.1	1.8	9.8
D10M5	35.3	42	43.8	15.95	4.11	5.3	5.5	6.1	3.6	9.8
D5M10	29.5	39.2	41.3	24.74	5.08	5.1	5.2	5.5	1.9	5.5
D10M10	30.3	37.5	40.5	19.20	7.41	5.1	5.3	5.4	3.8	1.9
	30.9	38.5	40.22	19.68	4.16	5.11	5.25	5.70	2.76	7.65

Hebhoub et al. (2011), stated that seventy percent of the marble considered as a waste during the process of mining and preparation of the materials. This huge amount of waste should be used in one way or another, to reduce the negative environmental impact of the marble. As the marble consumption grows and consequently, the production of marble increases, which in turn, decreases the strategic natural resources from the raw material. This waste is of alarming danger to the environment and require large areas of well-designed and prepared landfills.

The usage of ceramic waste has significant importance economically and environmentally; especially that it is considered as a safe disposal method for this type of construction wastes. Raval et al. (2013), tested the compressive strength and tensile strength of ceramic concrete concluded that, by increasing the replacement percentage of cement by ceramic powder up to 30% by weight, the compressive strength of the M20 grade concrete also increase. Any increase above this percentage in the cement replacement is accompanied by decrease in the strength. The 30% replacement for cement by ceramic powder gives 22.98 MPa for the compressive strength, and the cost of the cement reduced by 12.67% in the M20 grade concrete. Table 3, shows the design mix weights and volumes according to the Indian standard methods (IS 10262-2009), according to the same standards all the test samples were prepared. Table 4, identifies the design mix and the percentage of replacement for the cement in the concrete specimens.

This makes the concrete more economic without compromising its strength than the standard concrete and makes the concrete with ceramic powder used; technically and economically feasible and viable.

Table 3. Concrete mix design for cement replacement Raval et al. (2013)

	Water (Liter)	Cement (kg/m^3)	F.A. (kg/m^3)	C.A. (kg/m^3)		Chemical admixture
				20 mm	10 mm	
By Weight (kg)	169.3	325.5	730.2	759.2	506.1	2
By Volume (m^3)	0.52	1	1.8	2.35	1.49	–

Table 4. Different percentage of replacement Raval et al. (2013)

Number	Concrete type	Replacement percentage for cement by ceramic waste
1	A0	Standard concrete
2	A1	10% replacement
3	A2	20% replacement
4	A3	30% replacement
5	A4	40% replacement
6	A5	50% replacement

Giridhar et al. (2015), designed the concrete mix according to the Indian standard methods (IS 10262-2009), the mix proportions were as follows (0.48:1:1.53:2.88) - (Water: Cement: Fine Aggregate: Coarse Aggregate), as shown in Table 5. The mix design was done by weight for better concrete workability. As stated, the increase in ceramic waste coarse aggregate replacement in concrete, decreases the compressive strength values. At 100% replacement the compressive strength was found to be 32.15 MPa, which is more than the design target value –mean value- of the M20 grade concrete. Therefore, the 100% replacement concrete was assigned to be used as M20 grade concrete. The 40% replacement also gives a good compressive strength with loss for only 5.6%, which is marginal and could be tolerated. This ends up by the validity of using 40% replacement concrete to be used safely without taking into consideration the amount of loss in the compressive strength as it is not a significant value.

Split tensile strength results showed that the increase in the replacement percentage is accompanied by decrease in the tensile strength values. The results indicated that the decrease in the tensile strength is only 6.2% up to 20% replacement, which is a minor loss in the tensile strength. As a result, it was recommended that the usage of the 20% replacement concrete could be used safely instead of M20 grade concrete from the tensile strength tests.

Daniyal and Ahmad (2015), studied the fresh and hardened concrete properties for the replacement percentages from 0 to 50%-replacement of coarse aggregate by ceramic tiles aggregate-. The concrete mix was done according to the ratio (1:1.5:3) as shown in Table 6, and all the performed tests were according to the IS 516-1959. Slump,

Table 5. Concrete mix design with different replacement percentages Giridhar et al. (2015)

Number	Percentage of replacement	W/C ratio	Cement (kg/m^3)	FA (kg)	CA (kg)	Ceramic Aggregate (kg)	Water (Liter)
1	0	0.48	383	586	1103	0	183.8
2	20	0.48	383	586	882	221	183.8
3	40	0.48	383	586	662	441	183.8
4	60	0.48	383	586	441	662	183.8
5	80	0.48	383	586	221	882	183.8
6	100	0.48	383	586	0	1103	183.8

compressive strength, unit weight, and flexural strength were measured and the following results are obtained. For the fresh concrete, by increasing the replacement percentage the concrete workability decreases for all the mixtures by increasing the ceramic tile waste percentage. The concrete mass density decreases by increasing the water to cement ratio, also the density of concrete decreased by increasing the ceramic tile waste content. For the hardened concrete properties, regarding the concrete compressive strength, the concrete compressive strength increased by increasing the percentage of replacement up to certain limits, for example the 20% increase in the compressive strength for concrete with water to cement ratio of 0.4, 30% for concrete with water to cement ratio 0.5, and 40% for concrete with water to cement ratio of 0.6. The maximum compressive strength gained for concrete C5-10–10% replacement-. The increase in flexural strength was 32.2% higher than the flexural strength of the normal concrete, which means that the use of ceramic tile waste increases the flexural strength of concrete with considerable value. The use of ceramic tile waste in concrete has a useful effect, so as the usage of the ceramic waste in concrete enhance its properties, increase the compressive strength, the flexural strength, and also decrease its unit weight. The optimum use of ceramic tiles waste was found to be within the range of 10 to 30% of replacement.

Hemanth (2015), experimental program stated that the use of ceramic powder as replacement for fine aggregate, and the use of ceramic wastes as replacement of coarse aggregate enhanced the compressive strength by using 20% and 10% respectively. In case using both fine and coarse aggregate replacements simultaneously the compressive strength values increased in all cases. However, the optimum mix can be obtained by replacement of fine aggregate by 20% and the replacement of coarse aggregate by 10%. Also, by increasing the powder percentage, the workability of concrete increased, which means that the ceramic powder can be used in producing RMC "Ready Mix Concrete". By using the ceramic as a replacement for the coarse aggregate minor improvements to the workability occurred. Regarding the concrete mix design, all the mixes were done according to the IS 10262:2009. Nine different mixes were done as shown if the following Table 7, showing the different percentages of replacement for the fine aggregate and coarse aggregate.

Table 6. Concrete mix design coarse aggregate replacement with different water-to-cement ratios Daniyal and Ahmad (2015)

Group	W/C ratio	Cement (kg/m^3)	Water (kg/m^3)	FA (kg/m^3)	CA (kg/m^3)	Ceramic coarse aggregate (kg/m^3)
C4-0	**0.4**	300	120	450	900	0
C4-10		300	120	450	810	90
C4-20		300	120	450	720	180
C4-30		300	120	450	630	270
C4-40		300	120	450	540	360
C4-50		300	120	450	450	450
C5-0	**0.5**	300	150	450	900	0
C5-10		300	150	450	810	90
C5-20		300	150	450	720	180
C5-30		300	150	450	630	270
C5-40		300	150	450	540	360
C5-50		300	150	450	450	450
C6-0	**0.6**	300	180	450	900	0
C6-10		300	180	450	810	90
C6-20		300	180	450	720	180
C6-30		300	180	450	630	270
C6-40		300	180	450	540	360
C6-50		300	180	450	450	450

Shruthi et al. (2016), prepared a concrete mix –M20 grade- according to the IS 10262-2009, as shown in Table 8. The tests done on the specimens for the hardened concrete characteristics stated that, by the increase in the population the increase in the construction wastes also increase, which proves that the research on the usage of the construction wastes is a very important topic. The main concern of this study is to use the tiles wastes that come from the demolition of buildings, as partial replacement for aggregate in concrete mixes. The use of tiles in concrete has positive effects on the environment, also it helps in saving cost and non-renewable resources –natural aggregate–. The tile aggregate is cheaper than the natural aggregate. After all the experiments done on using tiles as a replacement for aggregate by certain amount of replacement the results are as follows, based on the compressive strength, split tensile strength tests. The maximum compressive strength was obtained from 30% replacement of natural aggregate by ceramic tile aggregate and the maximum split tensile strength was obtained from 30% replacement.

Table 7. Different percentage of replacement for fine aggregate and coarse aggregate Hemanth (2015)

Mix	Fine aggregate (%)		Coarse aggregate (%)	
	Sand	Tiles powder	C.A	Crushed tiles
A0	100	0	100	0
A1	100	0	90	10
A2	100	0	80	20
A3	90	10	100	0
A4	80	20	100	0
A5	90	10	90	10
A6	90	10	80	20
A7	80	20	90	10
A8	80	20	80	20

Table 8. Concrete mix design for different percentages of coarse aggregate replacement Shruthi et al. (2016)

Percentage of replacement of CTA	W/C ratio	Cement (kg/m^3)	Water (kg/m^3)	FA (kg/m^3)	CA (kg/m^3)	Ceramic coarse aggregate (kg/m^3)
0%	0.5	383.2	205.53	721.99	1099.66	NA
10%		383.2	220.79	721.99	989.7	94.7
20%		383.2	236.06	721.99	879.73	189.4
30%		383.2	251.33	721.99	769.76	284.1

Awoyera et al. (2016), focused in their study on the validity of using ceramic as a replacement for fine and coarse aggregate in concrete mix design. The concrete mix designed according to the British standards BS8110, as shown in Table 9 the concrete mix proportions. The results showed that the workability values for the ceramic waste aggregate concrete is better than the control concrete, for both coarse and fine aggregate replacements. The workability ranged between medium and high workability. And the mechanical properties of the ceramic waste aggregate (CWA) concrete enhanced, as the highest compressive strength and the highest split tensile strength were achieved from the 100% replacement of the natural aggregate by ceramic for both coarse and fine aggregate separately. The enhancement in the mechanical properties increase by the increase in the percentage of replacement, also within the scope of this study the ceramic waste aggregate concrete is suitable for construction. And also, if the strength is the main concern in the concrete mix design, the ceramic waste aggregate concrete is more efficient than the traditional concrete.

Table 9. Concrete mix for fine aggregate and coarse aggregate replacement Awoyera et al. (2016)

Mix	Cement (kg/m^3)	Fine aggregate (kg/m^3)		Coarse aggregate (kg/m^3)	
		Natural F.A.	CFA	Natural C.A.	CCA
CC	92	184	0	368	0
CFA-25	92	138	46	368	0
CFA-50	92	92	92	368	0
CFA-75	92	46	138	368	0
CFA-100	92	0	184	368	0
CCA-25	92	184	0	276	92
CCA-50	92	184	0	184	184
CCA-75	92	184	0	92	276
CCA-100	92	184	0	0	368

Adekunle et al. (2017), stated that concrete mixtures that uses ceramic waste is more sustainable and it uses the ceramic tiles as replacement for the aggregate in concrete with different replacement percentages making a new mixture with new properties – according to the ratio 1:2:4 Cement:FA:CA- as shown in Table 10 and 11. From the experimental results the compressive strength of concrete with 5% replacement can be used instead of the normal concrete as it deviates 11.37% from the strength of the control specimens after 28 days. Also, the deviation occurred due to the replacement of coarse aggregate by ceramic tile waste is 18.66% from the strength of the control specimens for compressive strength value corresponding to 25% replacement after 28 days. Regarding the ceramic tile fine replacement, it showed loss in concrete compressive strength up to 51.37% and 55.42% for 15% and 20% replacement respectively. These values are below the target strength of grade 20 concrete, so it is not recommended to use ceramic tiles as fine replacement in concrete with the mentioned percentages. For the coarse aggregate replacement, the concrete compressive strength reduced by using the replacement values 50% and 75% by 33.3% and 55.8% respectively corresponding to compressive strength values of 18.4 MPa and 12.2 MPa, which is also below the target strength values of grade 20 concrete. The optimum concrete mix design using ceramic tiles waste as a fine aggregate is 5% replacement, and the optimum concrete mix design using ceramic tiles as a coarse aggregate is 25% replacement. By increasing the percentage of replacement, the concrete mix for both fine and coarse aggregate replacement will not meet the target strength value of grade 20 concrete, which is from the code requirements as per BS 1881 part 4 (1997). Regarding the economical aspect, the use of ceramic tile as replacement of fine and coarse aggregate by 5% and 25% replacement respectively reduces the cost by 2.3% for every cubic meter of concrete.

Table 10. Concrete mix for fine aggregate replacement Adekunle et al. (2017)

Tile replacement (%)	W/C ratio	Cement (kg)	Sand (kg)	Granite (kg)
0	0.6	13.89	27.77	55.54
5		13.89	26.38	55.54
10		13.89	24.99	55.54
15		13.89	23.6	55.54
20		13.89	22.22	55.54

Table 11. Concrete mix for coarse aggregate replacement Adekunle et al. (2017)

Tile replacement (%)	W/C ratio	Cement (kg)	Sand (kg)	Granite (kg)
0	0.6	13.89	27.77	55.54
25		13.89	27.77	41.68
50		13.89	27.77	27.77
75		13.89	27.77	13.89

Sekar (2017), concluded that the usage of ceramic waste as a coarse aggregate replacement affect the concrete compressive strength, the optimum value in this study is 15% replacement. By experiments on the control beams and the beams using ceramic wastes, the ultimate load for the control beams is 610 KN, while the ultimate load for the 15% replacement is 600 KN, using the percentage of replacement shown in the following Table 12. The value of the ultimate load started to decrease by increasing the percentage of coarse aggregate by ceramic waste aggregate, as the ultimate load reached 565 KN and 550 KN for 30% and 45% replacement values respectively. So, the optimum percentage of replacement of coarse aggregate by ceramic is 15%.

Table 12. Concrete mix percentages of replacement Sekar (2017)

Mix	Percentage of replacement
C-15	15
C-30	30
C-45	45

Not only ceramic and marble; introducing RCA to the concrete is of main concern in research environmentally and economically. Concrete produced from the demolition of building and construction wastes contributes by 70% from the total amount of construction wastes, 14% for Asphalt concrete, 3% asphalt shingles, 2% brick and clay tiles, 1%

steel, 3% dry wall and plasters, 7% wood (USEPA 2016). This significant amount of concrete gives an indication on the importance of recycling this type of construction wastes and reusing it in the field of construction and especially in the concrete manufacture which is the second most consumed material (Ravindra et al. 2019).

The concrete compressive strength is the key characteristic property of concrete that is used to evaluate the concrete properties, quality and performance; as it has a significant correlation with other mechanical properties of concrete like tensile strength and flexural strength and also has a correlation with the concrete durability and its ability to resist abrasion, chemical penetration, and permeability. According to literature on the RCA collected from 79 publications used 977 different concrete mixes; the replacement of natural aggregates by RCA could not be studied just by studying the concrete mechanical properties; as the RCA concrete compressive strength at 28 days varied from a maximum and minimum values of 1.35 and 0.5 times the NA concrete compressive strength. This proves that the mechanical properties are not the only indication for the applicability of using such type of concrete (Ravindra et al. 2019).

The compressive strength decreases by increasing the replacement percentage of the NA by RCA. (Teranishi et al. 1998; Dhir et al. 1999; Limbachiya 2004; Etxeberria et al. 2007; Yang et al. 2008; Akbarnezhad et al. 2011; Fan et al. 2016). The concrete strength is directly related to the water to cement ratio, cement type, cement content, grain sizes and sources of aggregates used, the properties of the natural aggregates or recycled aggregates, and many other factors. All these factors should be studied comprehensively and correlated to the concrete compressive strength to determine the RCA concrete behavior (Ravindra et al. 2019).

The previous researches indicated different values for the optimum design of the concrete mixes for the coarse aggregate, fine aggregate, and cement replacement. However, the hardened concrete properties were not addressed in a comprehensive way, which means that the determination of the hardened concrete properties under the same conditions and the same materials to be used is not possible. In this research all the concrete mixes share the same materials and surrounding conditions to compare between the different characteristics fairly and without any change in the external factors that may affect the results and comparing it together.

4   Methodology

The concrete mix was designed according to the ACI code and the European standards. Fresh and hardened concrete properties were tested and analyzed to be compared by the traditional concrete. The required concrete strength is 35 MPa using natural aggregates, cement, and water to produce traditional normal concrete. Water to cement ratio was fixed for all the concrete mixes and equals to 0.48. the percentages of replacements of natural coarse aggregates by recycled concrete was 0%, 25%, 50%, 75%, and 100% -by volume- to be tested after 7, 28, and 90 days. The sustainable recycled aggregate concrete was designed to replace natural coarse aggregates by recycled concrete. Aggregates replacement was done by volume for different grain sizes. Using different sizes of aggregates by different percentages has a great impact on the enhancement of concrete fresh and hardened properties.

All aggregates used was oven dried till reaching constant mass; which means that the water content of the used aggregates equals to zero. Aggregates of different sizes were separated and sieves to remove fine marble dust less than 1 mm. the curing process for the fresh concrete specimens were done at room temperature (23 °C), and then demolded and transferred to water during the whole curing period. Specimens tested after 7, 28 and 90 days after measuring the exact dimensions for each type of specimens and calculate the exact force and stress in MPa. The machine used and the setup of the testing shown in Fig. 1.

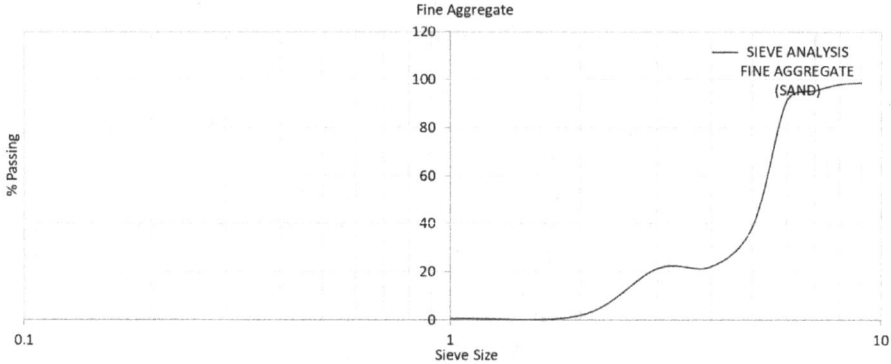

Fig. 1. Fine aggregate sieve analysis

After testing different types of sustainable concrete, a mathematical model was established and validated to estimate the mechanical properties of concrete taking into account the type of aggregates used and its properties. This mathematical model leads to an equation that could be used to determine the different mechanical properties compressive, split tensile and flexural strength; this equation involves some fresh and hardened concrete properties and the properties of the aggregates used.

5 Materials Used

Cement
Cement grade used is 42.5R, the initial setting time is 180 min and the final setting time is 280 min. The compressive strength of the cement after 2 days and 28 days is 34 MPa and 61.5 MPa, respectively.

Sand
Sand with grain size 0–2 mm is used in the concrete mix design with the following sieve analysis curves as shown in Fig. 2. The moisture content of sand is 0.1%.

Coarse Aggregates
Coarse aggregates, natural and recycled, with sizes from 2 mm to 16 mm are used with sieve analysis grading curves shown in Fig. 3. The used natural aggregates are well sieved, washed, and dried. recycled aggregate was used after being well dried until

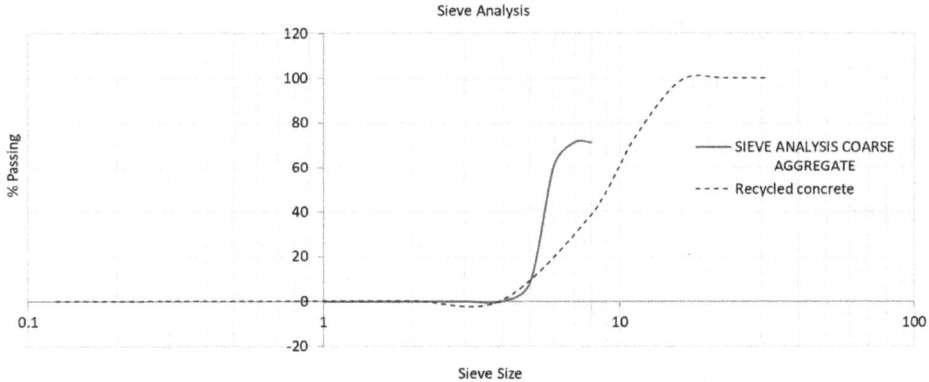

Fig. 2. Coarse aggregate sieve analysis (natural and recycled concrete aggregates)

constant mass. RCA abrasion, durability due to impact, moisture content and absorption were tested and found to be 30%, 25.6, 0 and 3.47 respectively. Same properties were tested for the natural aggregates to be able to compare the material properties and its effect on the green concrete behavior and found to be 41.1%, 30.6, 0.1 and 1.52 respectively.

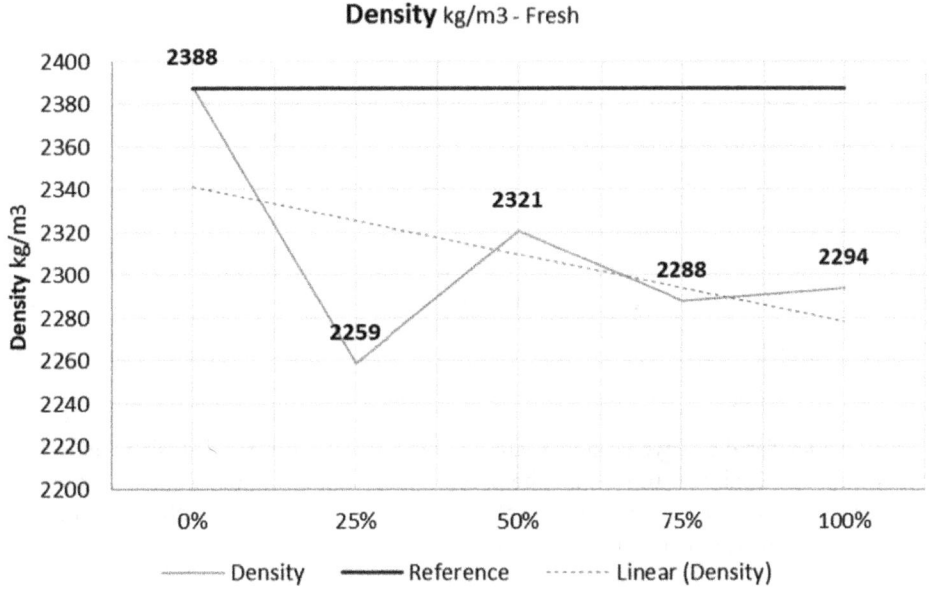

Fig. 3. Density of fresh concrete

Water

The properties of water as specified in the ACI, potable water should be used in concrete mixing with the accurate percentage and weight according to the design of the concrete mix.

6 Results and Discussion

Mechanical properties of concrete were tested using cubes (100 * 100 * 100 mm), cylinders (100 * 200 mm) and prisms (75 * 75 * 280 mm) to test concrete compressive strength, split tensile strength and flexural strength respectively.

Density
Density of fresh concrete shown in Fig. 4 slightly decreased by increasing the percentage of coarse aggregate in the concrete mix. The concrete density compared to the reference normal concrete decreased to reach 94.62%, 97.21%, 95.83% and 96.08% for 25%, 50%, 75% and 100% replacement of coarse aggregates by recycled aggregates; respectively.

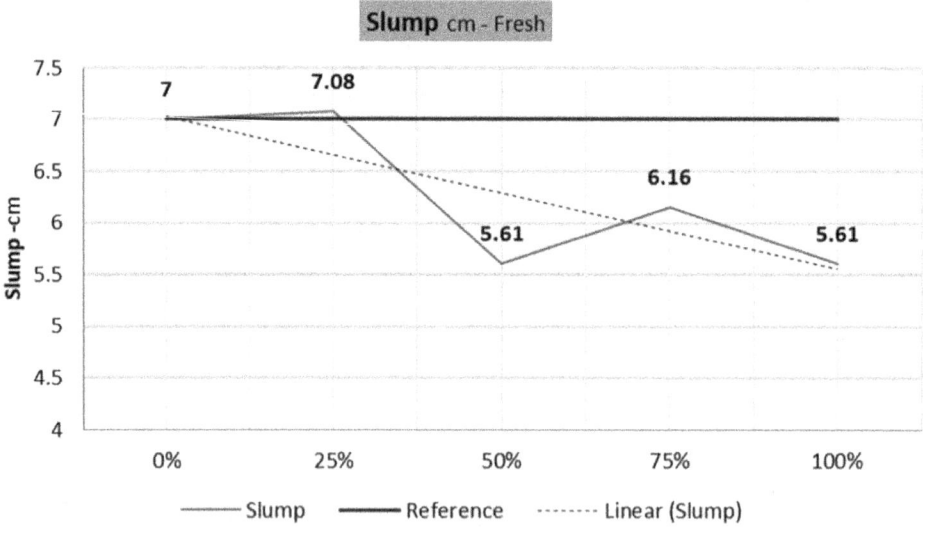

Fig. 4. Fresh concrete slump values

Slump
Slump values for the recycled aggregate concrete are all within the design range from 5 cm to 7 cm, as shown in Fig. 5. The slump values started to decrease by increasing the percentage of coarse aggregate replacement by recycled aggregate after the 25% replacement concrete which recorded slump value of 7.08 cm.

Shock Table
Shock table tests were performed on the recycled aggregate concrete to determine the quality of concrete concerning its consistency, as shown in Fig. 6, cohesiveness and its proneness to segregation. The reference value for the normal concrete was 40.5 cm and a slight decreasing behavior in the shock table test values was accompanied by increasing the percentage of replacement of natural aggregates with recycled aggregate. The minimum value was 37.5 cm and the maximum value was 41.5 cm which corresponds

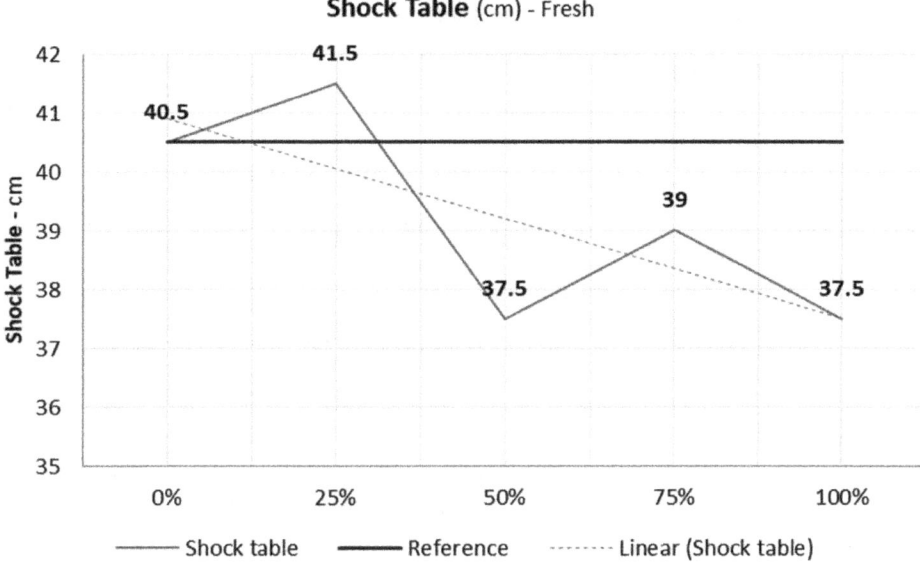

Fig. 5. Shock table values (measuring workability)

to the 50% and 25% replacement values, respectively. These values indicate concrete of class F2 (Plastic concrete), according to the European standards in the German index (DIN EN 206-1/DIN 1045-2); Table 13.

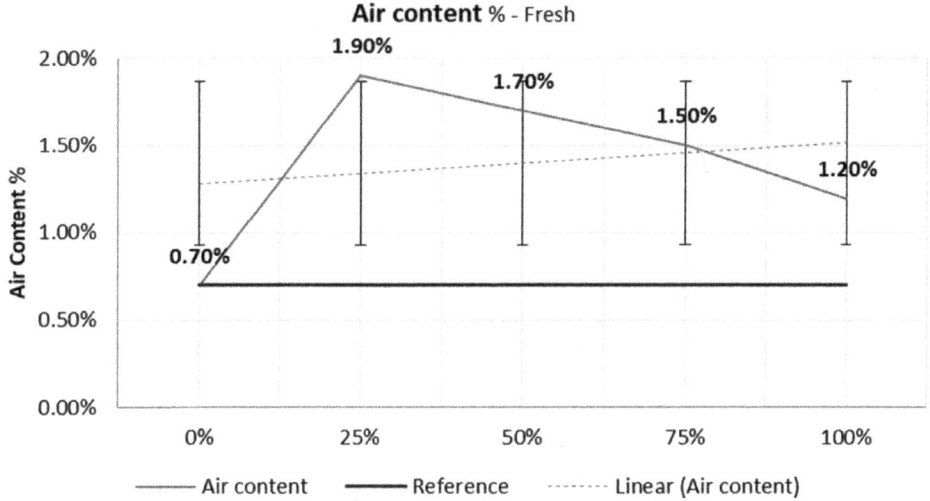

Fig. 6. Air content values for fresh concrete

Table 13. Workability classes as per the European standards

Grade	Slump values (mm)	Consistency range
F1	<340	Stiff
F2	350 to 410	Plastic
F3	420 to 480	Soft
F4	490 to 550	Very soft
F5	560 to 620	Very flowable
F6	≥630	Hardworking

Air Content

Air content values, as shown in Fig. 7, indicate all values are less than 2%, which is the value indicated by the requirements of ACI 318 match the recommended total air content values of ACI 210.2R, ACI 211.1, ACI 345, and ASTM C 94.

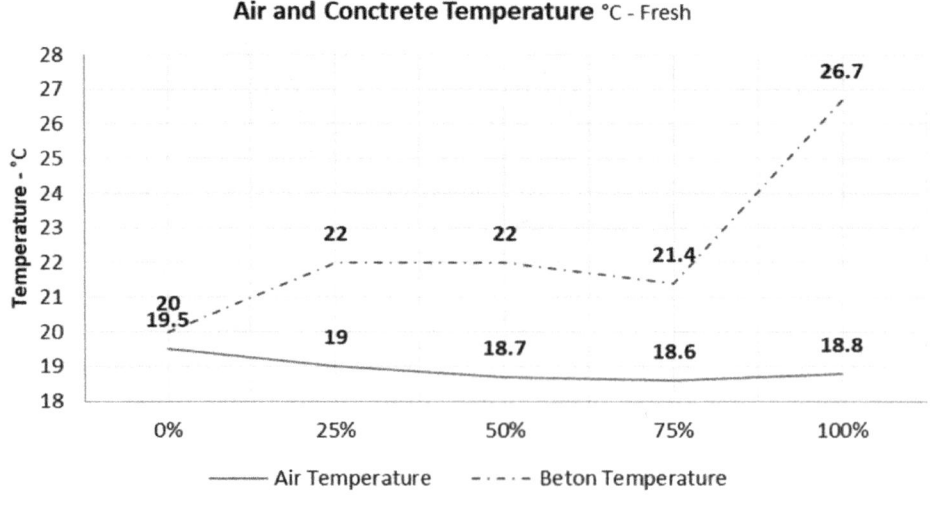

Fig. 7. Air and Concrete Temperature

Temperature

Concrete temperature had the same behavior of air temperature which indicates a strong relation between both temperatures; except for the 100% replacement of coarse aggregates by recycled aggregates which indicated an obvious increase in the concrete behavior as shown in Fig. 8.

Concrete Compressive Strength

As shown in Fig. 9 the average concrete compressive strength increased by increasing the percentage of recycled concrete starting from 100% which is the reference normal

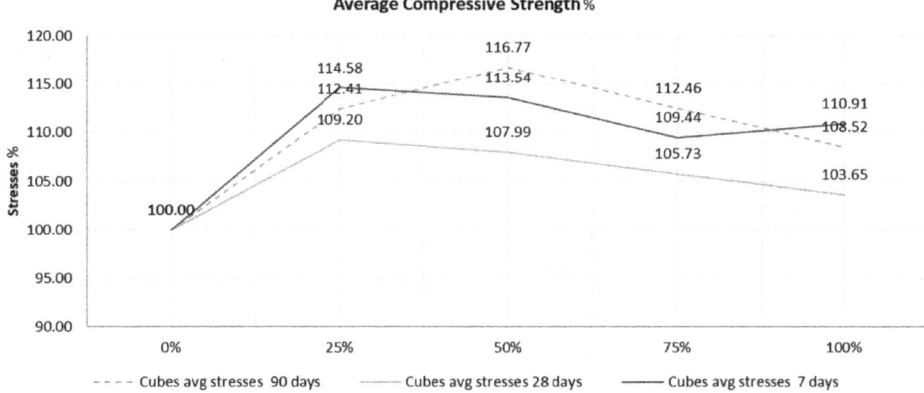

Fig. 8. Compressive Strength after different aging

concrete, then the compressive strength after the age of seven days increased to reach 114.58%, 113.54%, 109.44% and 110.91% for 25%, 50%, 75% and 100% replacement of coarse aggregates by recycled aggregates; respectively. After 28 days the concrete compressive strength improved to reach 109.2%, 107.99%, 105.73% and 103.65% for 25%, 50%, 75% and 100% replacement of coarse aggregates by recycled aggregates; respectively. the compressive strength after age of 90 days, reached an increase by 9% at the full replacement of the natural aggregates by the RCA when compared to the traditional concrete compressive strength at the same age. Traditional concrete compressive strength value after 90 days is 100%, reaching 112.41%, 116.77% and 112.46%, 108.52% for replacement percentages equivalent to 25%, 50%, 75% and 100% respectively.

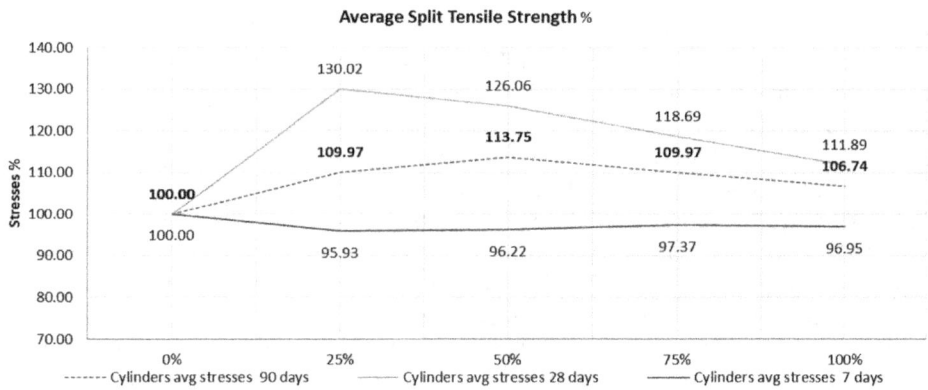

Fig. 9. Split tensile Strength after different aging

Concrete Split Tensile Strength

Figure 10 shows the concrete split tensile strength values which decreased slightly after the age of seven days and increased at the age of twenty-eight days. The concrete split tensile strength after 7 days decreased by approximately 4%, 4%, 3% and 3% for 25%, 50%, 75% and 100% replacement of coarse aggregates by recycled aggregates; respectively. An increase in the split tensile strength occurred after 28 days by 30%, 26%, 19% and 12% for 25%, 50%, 75% and 100% replacement of coarse aggregates by recycled aggregates; respectively. By increasing the recycled concrete aggregates in concrete after 90 days the concrete flexural strength increased reaching 109.97%, 113.75%, 109.97% and 106.74% for percentages of replacements 25%, 50%, 75% and 100% respectively. The value of seven percent gain strength was recorded for the 100% replacement when compared to the traditional concrete.

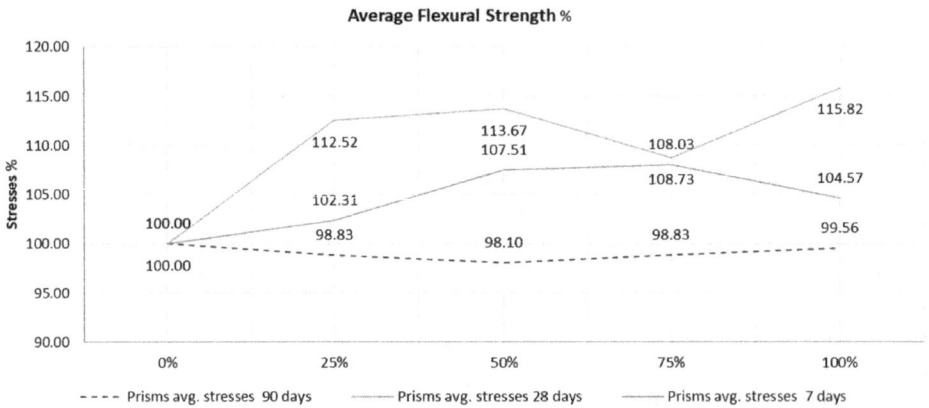

Fig. 10. Flexural Strength after different aging

Concrete Flexural Strength
The average flexural strength shown in Fig. 11 increased by increasing the percentages of replacement of coarse aggregates by recycled concrete aggregates after aging of seven and twenty-eight days. At the age of seven days the flexural strength increased from 100% to 102.31%, 107.51%, 108.03% and 104.57% for 25%, 50%, 75% and 100% replacement of coarse aggregates by recycled aggregates; respectively. After 28 days the flexural strength gained higher values of strength reaching 112.52%, 113.67%, 108.73% and 115.87% for 25%, 50%, 75% and 100% replacement of coarse aggregates by recycled aggregates; respectively. The flexural strength of concrete after age of 90 days which decreased from 100% for the traditional concrete reaching 98.93%, 98.1%, 98.83% and 99.56% for 25%, 50%, 75% and 100% replacement values for natural aggregates by RCA.

Microstructure Analysis
All the microstructure analysis tests showed a perfect bond between the concrete particles especially at the interfacial transition zone (ITZ), nonoccurrence of micropores, air voids and microcracks. The concrete microstructure images showed homogenous paste as shown in the following Figs. 11, 12, 13, 14 and 15.

Fig. 11. Microstructure analysis for reference concrete

Fig. 12. Microstructure analysis for RCA concrete 25%

Fig. 13. Microstructure analysis for RCA concrete 50%

Fig. 14. Microstructure analysis for RCA concrete 75%

Resistance to Abrasion

The concrete resistance to abrasion was tested according to the DIN 52108 standards, for both the normal concrete and the RCA concrete and the results showed a decrease in the resistance to abrasion by increasing the percentage of the RCA in concrete till

Fig. 15. Microstructure analysis for RCA concrete 100%

reaching 50% replacement. The behavior of the RCA after the 50% replacement showed an increase in the concrete abrasion resistance for the 75% and 100% replacement. However, the 100% replacement of natural aggregates by RCA showed a decrease in the resistance due to abrasion by 13% when compared to the normal concrete, as shown in Fig. 16.

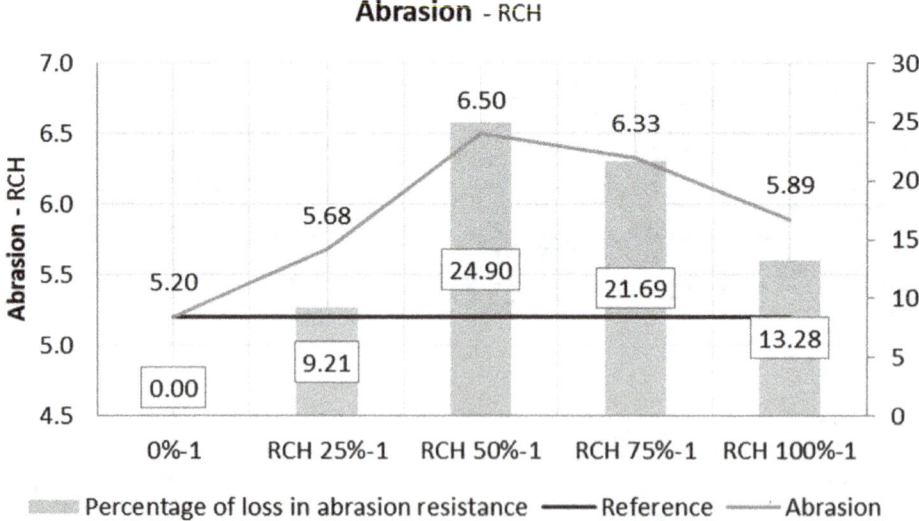

Fig. 16. Abrasion resistance for RCA concrete

7   Failure Modes

All failure modes were observed and photographed; compared to the traditional concrete failure mode and were found similar in all percentages of replacement. The cracking and failure behavior of specimens were observed to determine the ductile behavior of the manufactured concrete. After testing all the failure modes and the cracking behavior of the different kinds of specimens -cubes, cylinders, and prisms- the concrete behavior was determined as ductile -including the reference specimens- compared to other specimens from other batches.

8 Mathematical Model and Design Equation

A design equation concluded from the green concrete fresh and hardened properties Eq. (1), estimates the average concrete compressive strength after seven days of curing. The design equation includes the fresh concrete density, concrete air content at fresh state, slump values, air temperature at the time of mixing, and the concrete temperature at mixing time. The aggregates hardness factor could be selected from the Tables 14 a, b, and c.

Table 14. (a, b & c) Aggregates properties that determines the concrete mechanical properties and factors in the green concrete design equation first mathematical model

Shape		Hardness tests		Cumulative	
Aggregates Roughness Factor		Aggregates Hardness Factor		Aggregates Hardness Factor	
Normal Aggregate	1	Normal Aggregate	1	Normal Aggregate	1
Marble	1	Marble	0.95	Marble	0.95
Ceramic	1.45	Ceramic	0.35	Ceramic	0.5
Recycled Concrete	1.55	Recycled Concrete	0.78	Recycled Concrete	1.2

Where; density in kg/cm^3, Air content in %, Slump and shock table in centimeters, Air and concrete temperatures in °C, Aggregates Hardness Factor from Table 14-c.

$$\text{Average Concrete Compressive Strength (7 days)} = ((\text{Density} + \text{Air content} + \text{Slump} + \text{shock table} + \text{Air Temperature} + \text{Concrete Temperature}) * \text{AHF})/55.2 \tag{1}$$

9 Conclusions

Using recycled concrete aggregates in sustainable green concrete manufacture enhanced the fresh and hardened concrete properties significantly especially at the full replacement of natural aggregates by the recycled concrete aggregates; after testing different percentages of replacements -25%, 50%, 75% and 100%- in green concrete and comparing the results by the reference normal concrete. The benefit from this full replacement of coarse aggregates by sustainable recycled concrete aggregates has environmental and economical benefits in addition to the improvement of the concrete characteristics under the same water to cement ratio and external factors with high quality control and restrictions during concrete mixing, pouring curing and testing.

The tests done on the concrete mechanical and microstructure properties proved the usability of RCA-concrete for high strength structural elements reaching 66.2 MPa, 3.96 MPa and 6.8 MPa for concrete compressive, split tensile and flexural strength after 90 days respectively.

References

Adams, M.P.: Alkali-silica reaction in concrete containing recycled concrete aggregates (Doctoral dissertation) (2012)

Adekunle, A.A., Abimbola, K.R., Familusi, A.O.: Utilization of construction waste tiles as a replacement for fine aggregates in concrete. Eng. Technol. Appl. Sci. Res. **7**(5), 1930–1933 (2017)

Akbarnezhad, A., Ong, K.C.G., Zhang, M.H., Tam, C.T., Foo, T.W.J.: Microwave-assisted beneficiation of recycled concrete aggregates. Constr. Build. Mater. **25**(8), 3469–3479 (2011)

Ananda Ramakrishna, K., Sateesh Babu, K., Guravaiah, T., Naveen, N., Sk, J.: Effect of waste ceramic tiles in partial replacement of coarse and fine aggregate of concrete. Int. Adv. Res. J. Sci. Eng. Technol. **2**(6), 13–16 (2015)

ASTM, C.: Standard test method for flexural strength of concrete (using simple beam with third-point loading). American Society for Testing and Materials, Philadelphia, PA (1999)

Basri, H.B., Mannan, M.A., Zain, M.F.M.: Concrete using waste oil palm shells as aggregate. Cem. Concr. Res. **29**(4), 619–622 (1999)

Binici, H.: Effect of crushed ceramic and basaltic pumice as fine aggregates on concrete mortars properties. Constr. Build. Mater. **21**(6), 1191–1197 (2007)

Binici, H., Shah, T., Aksogan, O., Kaplan, H.: Durability of concrete made with granite and marble as recycle aggregates. J. Mater. Process. Technol. **208**(1–3), 299–308 (2008)

Code, E.: Egyptian Code of Practice for Concrete Structures, HBRC. Arabic, Cairo, Egypt (2007)

Daniyal, M., Ahmad, S.: Application of waste ceramic tile aggregates in concrete. Int. J. Innov. Res. Sci. Eng. Technol. **4**(12), 12808–12815 (2015)

Ergün, A.: Effects of the usage of diatomite and waste marble powder as partial replacement of cement on the mechanical properties of concrete. Constr. Build. Mater. **25**(2), 806–812 (2011)

Etxeberria, M., Vazquez, E., Mari, A., Barra, M.: Influence of amount of recycled coarse aggregates and production process on properties of recycled aggregate concrete. Cem. Concr. Res. **37**(5), 735–742 (2007)

Fan, C.C., Huang, R., Hwang, H., Chao, S.J.: Properties of concrete incorporating fine recycled aggregates from crushed concrete wastes. Constr. Build. Mater. **112**, 708–715 (2016)

Giridhar, V., Rao, H.S., Kumar, P.S.P.: influence of ceramic waste aggregate properties on strength of ceramic waste aggregate concrete. Int. J. Res. Eng. Technol. **4** (2015)

Hebhoub, H., Aoun, H., Belachia, M., Houari, H., Ghorbel, E.: Use of waste marble aggregates in concrete. Constr. Build. Mater. **25**(3), 1167–1171 (2011)

Heidelberg Cement, AG: Concrete technical data (2014)

Heidelberg Cement, AG: Concrete technical data (2017)

IS 10262-2009: Concrete mix proportioning—guidelines

Kavitha, B., Sundar, M.L.: Experimental study on partial replacement of coarse aggregate with ceramic tile wastes and cement with glass powder (2017)

Khalaf, F.M., DeVenny, A.S.: Recycling of demolished masonry rubble as coarse aggregate in concrete. J. Mater. Civ. Eng. **16**(4), 331–340 (2004)

Kou, S.C., Poon, C.S., Lam, L., Chan, D.: Hardened properties of recycled aggregate concrete prepared with fly ash. In: Limbachiya, M.C., Roberts, J.J. (eds.) Proceedings of the International Conference on Sustainable Waste Management and Recycling: Challenges and Opportunities, London, UK, pp. 189–197 (2004)

Kumar, P.S., Mannan, M.A., Kurian, V.J., Achuytha, H.: Investigation on the flexural behaviour of high-performance reinforced concrete beams using sandstone aggregates. Build. Environ. **42**(7), 2622–2629 (2007)

Manso, J.M., Polanco, J.A., Losanez, M., Gonzalez, J.J.: Durability of concrete made with EAF slag as aggregate. Cem. Concr. Compos. **28**(6), 528–534 (2006)

Medina, C., De Rojas, M.S., Frías, M.: Reuse of sanitary ceramic wastes as coarse aggregate in eco-efficient concretes. Cem. Concr. Compos. **34**(1), 48–54 (2012)

El-Hawary, M.T., Hanna, N.F., El-Nemr, A.M., Koenke, C.: Using construction wastes and recyclable materials in sustainable concrete manufacture. In: ICSBMCT 2020: 14 International Conference on Sustainable Building Materials and Construction Technologies at Barcelona, Spain (2020)

Nataraja, M.C., Nagaraj, T.S., Reddy, A.: Proportioning concrete mixes with quarry wastes. Cem. Concr. Aggreg. **23**(2), 81–87 (2001)

Nehdi, M., Khan, A.: Cementitious composites containing recycled tire rubber: an overview of engineering properties and potential applications. Cem. Concr. Aggreg. **23**(1), 3–10 (2001)

Obe, R.K.D., de Brito, J., Silva, R.V., Lye, C.Q.: Sustainable Construction Materials: Recycled Aggregates. Woodhead Publishing, Sawston (2019)

Palmquist, S.M., Jansen, D.C., Swan, C.W.: Compressive behavior of concrete with vitrified soil aggregate. J. Mater. Civ. Eng. **13**(5), 389–394 (2001)

Vanghiyan, R.: Workability by replacement of cement with waste glass. Int. J. Innov. Res. Sci. Eng. Technol. **3**(7) (2013)

Raval, A.D., Patel, I.N., Pitroda, J.: Ceramic waste: Effective replacement of cement for establishing sustainable concrete. Int. J. Eng. Trends Technol. (IJETT) **4**(6), 2324–2329 (2013)

Senthamarai, R.M., Manoharan, P.D.: Concrete with ceramic waste aggregate. Cem. Concr. Compos. **27**(9–10), 910–913 (2005)

Taj, S., Pasha, S.R.: Reuse of Ceramic Waste as Aggregate in Concrete (2016)

Teranishi, K., Dosho, Y., Narikawa, M., Kikuchi, M.: Application of recycled aggregate concrete for structural concrete. Part 3-production of recycled aggregate by real-scale plant and quality of recycled aggregate concrete. In: Sustainable Construction: Use of Recycled Concrete Aggregate: Proceedings of the International Symposium Organised by the Concrete Technology Unit, University of Dundee and held at the Department of Trade and Industry Conference Centre, London, UK, 11–12 November 1998, pp. 143–156. Thomas Telford Publishing (1998)

Tests, C.S.: ASTM C 39/C 39 M; test one set of two laboratory-cured specimens at 7 days and one set of two specimens at 28 days. a. Test one set of two field-cured specimens at, 7

USEPA: Construction and Demolition Debris Generation in the United States. U.S. Environment Protection Agency, Office of Resource Conservation and Recovery (2016). 23 p

Yang, K., Chung, H., Ashour, A.: Influence of type and replacement level of recycled aggregates on concrete properties. ACI Mater. J. **105**(3), 289–296 (2008)

Zimbili, O., Salim, W., Ndambuki, M.: A review on the usage of ceramic wastes in concrete production. Int. J. Civ. Archit. Struct. Constr. Eng. **8**, 91–95 (2014)

A Study to Evaluate the Energy Performance for the Brownfield Building at SVNIT, Surat, Gujarat

Krupesh A. Chauhan[1(✉)], Bhagyashri H. Sisode[1], Govind Pandey[2],
and Vinay Kumar Singh[2]

[1] SVNIT, Surat, India
kac@ced.svnit.ac.in, bhagyashrisisode14@gmail.com
[2] MMMUT, Gorakhpur, India
{gpce,vksce}@mmmut.ac.in

Abstract. Across the different parts of the world, the concept of Green Building is catching attention of the architects, urban & town planners, engineers and even the common people. According to the experts, Green buildings are the ones that benefit the environment as well as the human beings by minimum waste generation at all the stages of its life cycle while being cost effective, sustainable and providing comfort to the humans. The presence and behaviour of consumers in the buildings impact largely on energy utilization of various devices such as artificial lights, air conditioners, heaters, etc. Building's designed energy performance can be affected to one-third by unaware energy usage behaviour. To save the energy, the research team has carried out a pilot study in the Civil Engineering Department (CED), SVNIT, Surat by combining the concepts of Energy Performance and Value Engineering. All the lights in the passage of the department, for the both the floors, were replaced with the sensor lights with specifically designed automated fixtures by referring Energy Conservation Building Code (ECBC) 2017, Bureau of Energy Efficiency (BEE), MoP, GOI. The results obtained showed unmatchable tangible benefits and proved that by implementing the same at the institute level can prove to save considerable amount of cost, water used to produce electricity and also reduce the carbon foot prints. As per the analysis, the annual kW usage reduced to 54.02 kWh from 262.8 kWh, saving 79.44%. The resulting total annual saving is estimated to be 208.78 kWh. The outcome of this research will be useful for administration of the Institute. Furthermore, the Institute is demonstrating this pilot building for energy performance study to the networking institutes, students and faculty members as well as policy makers, planners, developers and end users.

Keywords: Energy performance · Green and brownfield building · Performance · Value engineering

1 Introduction

The Green Building, as defined by American Society for Testing and Materials (ASTM) International (2001) is, "A building that provides the specified building performance requirements while minimizing disturbance to and improving the functioning of local, regional, and global ecosystems both during and after its construction and specified service life". Furthermore, "A green building optimizes efficiencies in resource management and operational performance; and minimizes risks to human health and the environment".

Initially, the word brownfield was used in construction and development of a land that had a permanent structure. Presently, this term is used in numerous industries to launch a project based to engineer or rebuild a product from an existing one.

Value Engineering adds value and reduces the cost of providing customer satisfaction, especially in the long run. Value engineering can be applied in the structural, architectural and material components of the building to enhance cost-effectiveness (Janani et al. 2018). Energy performance is a measurement of the overall energy usage, as determined by the resources needed to provide construction operation for structures, construction machinery or building materials.

There are plethora of definitions for Intelligent building out which two definitions provided by Intelligent Building Institute (IBI), United States and European Intelligent Building Group (EIBG), United Kingdom. According to the definition provided by IBI, the building that provides a dynamic and cost-efficient environment by optimizing its four basic elements that includes structures, systems, services and management along with the interdependence amongst them, is the Intelligent Building. Whereas, according to EIBG, the Intelligent Building is the one generated by optimizing the effectiveness of the occupancy of the building thus allowing effective resource utilization with minimal life-time costs of hardware and facilities (Wigginton and Harris 2002 and Nguyen et al. 2013).

When the concepts of green building, brownfield, value engineering and energy performance are collaborated, the resulting structure is an intelligent building.

The research was carried out at Sardar Vallabhbhai National Institute of Technology, Surat (Established in 1961) which is one of the premier Institutes of National Importance and fully financed and controlled by the Government of India, New Delhi. It has experienced faculty and dynamic research scholars' team & office staff.

2 Need of Study

In the contemporary world like today's, when there are innumerable environmental issues, it is the need of the hour to plan and design energy efficient green building to fulfil the demand of the user along with conserving the environment. The energy consumption in buildings represents a large proportion of the total end consumption of energy. Lighting accounts for nearly a fifth of the world's electricity consumption. With the outpaced technological developments, it is possible to save the energy and make the building energy efficient. Nonetheless, when the modifications are made in the brownfield building by replacing the old fixtures with advanced ones, it is inevitable to know if the modifications are fruitful. The use of the artificial lights which is unavoidable. But, lot of electricity and thus cost can be saved when the old lights are replaced with the sensor lights. The evaluation of the usage of such lights in the brownfield buildings is a domain not addressed frequently. Ergo, this research is important and the outcome obtained through this research will be useful in many aspects.

3 Study Area

Surat, situated on the banks of river Tapi and the coast of the Arabian Sea, is in the Gujarat state of India. It is a high ranking industrial city of the country with a strong network of roads & flyovers. The campus of SVNIT, Surat, is located within the South West Zone of Surat city. The institute was initially established as Regional Engineering College (REC) in 1961 and was upgraded as a National Institute of Technology (NIT) with the status of 'Deemed University' in 2002. The map of Surat city and the master plan of SVNIT, Surat are shown Fig. 1 and 2 respectively.

   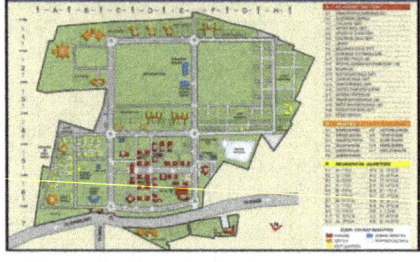

Fig. 1. Map of Surat   **Fig. 2.** Master plan of SVNIT, Surat

CED is one of the pioneering departments, since the commencement of the college in 1961. It is one of the most accessible buildings from the main gate of campus. The latitude and longitude of the CED of the SVNIT Campus are 21.1740°N and 72.7937°E. The total built up area of the building is 1970.61 m^2. It comprises of 13 rooms at the ground floor and 25 rooms at the first floor. Apart from four major faculty rooms, the building also houses a lecture hall, drawing hall, environmental laboratory, survey laboratory, administrative office and one common staff toilet on the ground floor. The first floor has 7 faculty rooms, 5 research scholar rooms, a seminar hall, a department library, girl's room, 2 toilets blocks and 5 laboratories.

4 Objective

The objective of this research paper is to evaluate the energy saved when the lights with sensors are installed.

5 Scope of Study

The scope of the study was limited to the Civil Engineering Department, SVNIT, Surat.

6 Methodology

Firstly, the Normal Lights were replaced with the Smart LED (Light Emitting Diode) Lights equipped with sensors in the corridor of the ground floor and first floor of CED, SVNIT, Surat. Along with this change, energy meters were attached to measure the consumption of energy of the normal light and smart LED tube lights, respectively. The readings from the energy meter were collected on a regular basis for one and a half month. The comparison was done while comparing the readings for different types of aforesaid lighting systems. Furthermore, to get more clear picture about the energy consumption on week days and weekends, separate evaluation was carried out.

7 Data Collection

The CED building is a double storied building with the main entrance facing towards the East direction. The building has a ground floor build up area of 968.48 sq. m and a first floor build up area of 958.58 m^2. The terrace area of the building is 43.55 m^2. The building has a length of 54.77 m (i.e. in the East-West direction) and a width of 18.33 m (i.e., in the North- South direction). The plan details of the ground floor and first floor of CED are shown in Fig. 3.

Fig. 3. Detailed plan of ground floor and first floor of CED, SVNIT, Surat.

After the installation of sensor lights, the readings were collected through the energy consumption meters for normal as well as smart lights. The Fig. 4 shows the initial stage of the energy meter with '0.00' reading in both the meters.

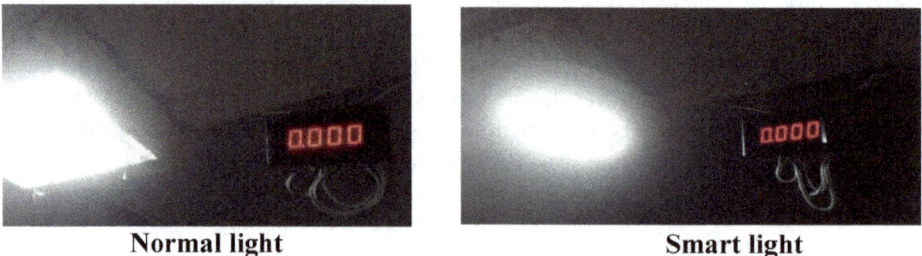

| Normal light | Smart light |

Fig. 4. Energy meter reading at initial stage for normal light and smart light

Likewise, the readings were taken everyday at the same time i.e. 18:00 h. After the first day, the readings were 0.638 kWh and 0.129 kWh respectively for the normal and smart lights as shown in the Fig. 5. Similarly, the data collected for 45 days in row. The readings on the 45th day were 27.038 kWh for the normal light and 5.559 kWh for the smart light.

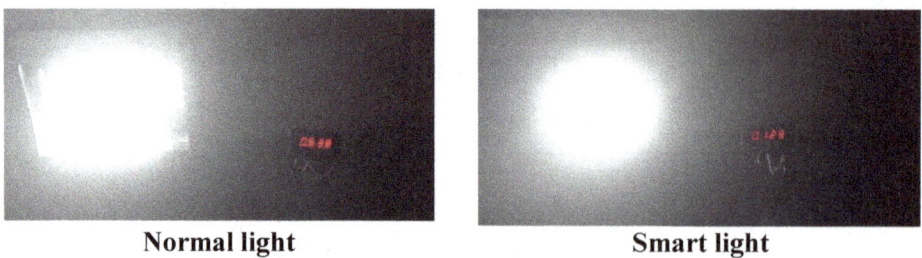

| Normal light | Smart light |

Fig. 5. Energy after 1 day (kWh)

8   Analysis

The smart surface lights installed in the corridors of the Civil Engineering Department are shown in the Fig. 6 and the sensor used in the light is shown in Fig. 7. The sensor is capable of sensing the daylight as well as the presence of the human beings.

The motion detector turns on the light based on movement. With this built in detector, light is automatically on when needed and off when not in need.

The Fig. 8 (a) shows that the light remains off during daytime even when human movement is detected. This is because of the daylight sensing. In this case, the ambient lux level is above the present daylight threshold. Figure 8 (b) shows that the light is triggered on by the detector with the movement and insufficient ambient lux level and

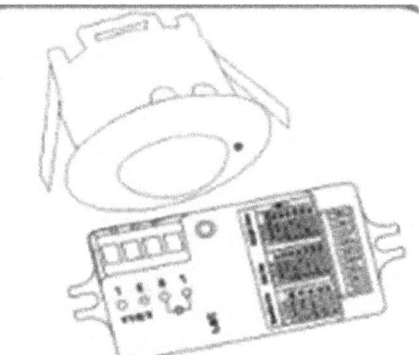

Fig. 6. Smart surface lights   **Fig. 7.** Sensors

the Fig. 8 (c) shows that even when there is insufficient ambient lux level, the light turns off automatically after the hold-time pre-set, if there is no more movement. The sensor used for such automation is hidden and hence any one cannot judge or cut off the wire to deactivate it. The inbuilt man movement sensor saves the power by 30% more.

The energy meter attached with the normal and smart lights, showed the cumulative reading. Those readings were collected and from them the daily consumption was found out by subtracting the reading of the previous day. The energy consumed on daily basis is shown in the Table 1. The table delineates the difference in the energy consumption clearly.

Furthermore, it can be seen that the least energy is consumed on weekends, especially on Sunday. Though there is holiday on Sunday, yet the department is open for the researchers and faculty members who want to visit department for their laboratory or research work. This is the reason why there is the consumption of energy even on Sunday. Nevertheless, it is minimum as compared to all other working days.

When the usage of six weekends was calculated, it was found out that 3.701 kWh was consumed by the normal lights and 0.867 kWh was consumed by the smart light and that of the weekdays in the duration of 45 days was 23.337 kWh and 4.692 kWh for normal and smart lights respectively.

These days, CFL is widely used. Table 2 shows the comparison of the normal light and the smart light. The figures in the table elucidates that the smart light is much more efficient comparatively because of the intelligent technology.

The results of the 45 days experiment are shown in the Table 3. After analysing the energy consumed for 45 days, the average energy consumed daily is 0.601 kWh and 0.124 kWh respectively for normal and smart lights. The annual estimation shows that 79.44% energy will be saved.

When looked upon the through Value Engineering aspect, the capital invested in installing the smart lights will be paid back within 4 to 8 months.

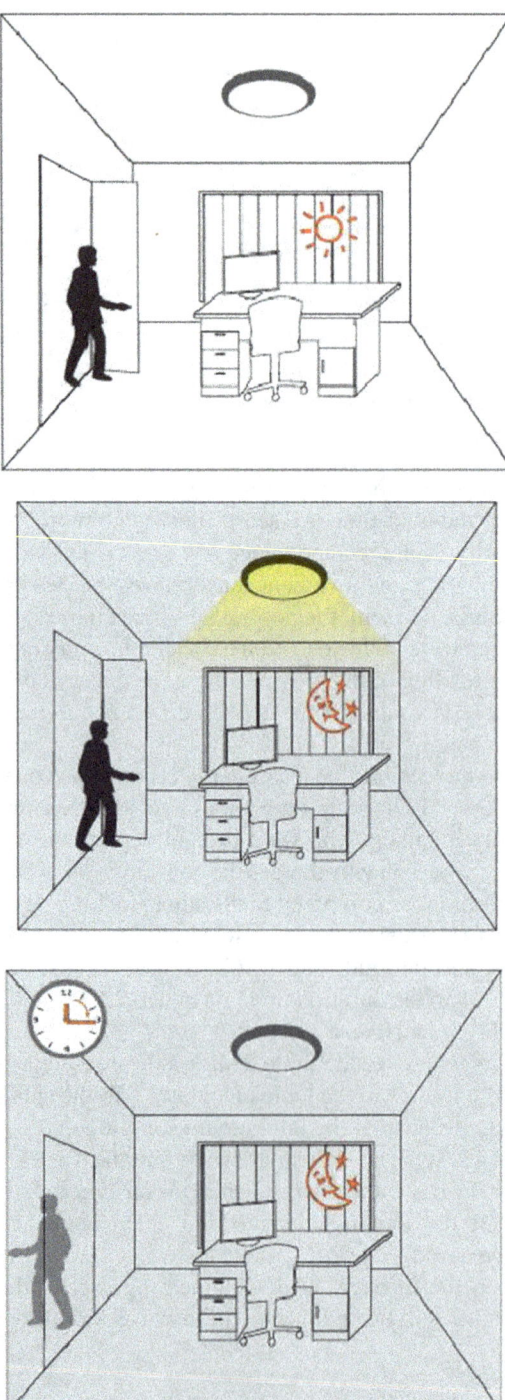

Fig. 8. (a) Behaviour of the smart light in the presence of the sunlight. (b) Behaviour of the smart light in the absence of the sunlight with the presence of human. (c) Behaviour of the smart light in the absence of the sunlight and human.

Table 1. Day wise energy meter readings and energy consumption

Sr. no.	Day	Cumulative energy meter readings		Daily energy consumption	
		Normal light	Smart light	Normal light	Smart light
1	Monday	0.638	0.129	0.638	0.129
2	Tuesday	1.372	0.281	0.734	0.152
3	Wednesday	2.067	0.422	0.695	0.141
4	Thursday	2.811	0.592	0.744	0.17
5	Friday	3.513	0.739	0.702	0.147
6	Saturday	3.952	0.831	0.439	0.092
7	Sunday	4.155	0.874	0.203	0.043
8	Monday	4.857	1.022	0.702	0.148
9	Tuesday	5.496	1.152	0.639	0.13
10	Wednesday	6.23	1.306	0.734	0.154
11	Thursday	6.986	1.464	0.756	0.158
12	Friday	7.675	1.608	0.689	0.144
13	Saturday	8.087	1.694	0.412	0.086
14	Sunday	8.286	1.736	0.199	0.042
15	Monday	8.986	1.883	0.7	0.147
16	Tuesday	9.64	2.02	0.654	0.137
17	Wednesday	10.37	2.173	0.73	0.153
18	Thursday	11.156	2.338	0.786	0.165
19	Friday	11.805	2.474	0.649	0.136
20	Saturday	12.207	2.558	0.402	0.084
21	Sunday	12.417	2.602	0.21	0.044
22	Monday	13.068	2.738	0.651	0.136
23	Tuesday	13.766	2.884	0.698	0.146
24	Wednesday	14.518	3.042	0.752	0.158
25	Thursday	15.317	3.209	0.799	0.167
26	Friday	15.952	3.342	0.635	0.133
27	Saturday	16.375	3.431	0.423	0.089
28	Sunday	16.61	3.48	0.235	0.049
29	Monday	17.295	3.624	0.685	0.144
30	Tuesday	17.954	3.762	0.659	0.138
31	Wednesday	18.663	3.911	0.709	0.149

(*continued*)

Table 1. (*continued*)

Sr. no.	Day	Cumulative energy meter readings		Daily energy consumption	
		Normal light	Smart light	Normal light	Smart light
32	Thursday	19.478	4.082	0.815	0.171
33	Friday	20.176	4.228	0.698	0.146
34	Saturday	20.571	4.311	0.395	0.083
35	Sunday	20.765	4.352	0.194	0.041
36	Monday	21.405	4.486	0.64	0.134
37	Tuesday	22.16	4.644	0.755	0.158
38	Wednesday	22.86	4.791	0.7	0.147
39	Thursday	23.574	4.941	0.714	0.15
40	Friday	24.296	5.002	0.722	0.061
41	Saturday	24.698	5.176	0.402	0.174
42	Sunday	24.885	5.216	0.187	0.04
43	Monday	25.608	5.36	0.723	0.144
44	Tuesday	26.264	5.506	0.656	0.146
45	Wednesday	27.038	5.559	0.774	0.053

Table 2. Comparison – normal light (compact fluorescent lamp - CFL) vs smart light

Considerations	Normal CFL light	Smart LED light
Fixture watts:	30 W	20 W
Annual hrs. operation: (24 h per day)	8760 h	8760 h
Total operating power	30 W	20 W
Output lumens	2050 lm	1950 lm
Light efficiency	68 lm/W	98 lm/W
Light output after 10 k hr operation	75%	98%
Rated lamp life	10,000 h	20,000 h

Table 3. Results of 45 days experiment

Considerations	Normal LED TL	Smart LED TL
Operation hours	45 Days × 24 h = 1080 h	45 Days × 24 h = 1080 h
45 days energy usage	27.038 kWh	5.559 kWh
Average energy consumed daily	0.601 kWh	0.124 kWh
Estimated annual energy consumption	262.8 kWh	54.02 kWh
Annual savings:	–	208.78 kWh
Savings percentage:	–	79.44%

9 Conclusion

This paper elucidates the research in the area of energy conservation leading a step to redevelop a brownfield building into an intelligent building. Specifically, the paper delineates the benefits of replacing the traditional or the normal artificial lighting system with the intelligent or the smart lighting system with sensors. The outpaced technology has unmatchable benefits that one has to grab. By the installation of the smart lights instead of the normal lights can save up to 79.44% of energy annually in CED, SVNIT, Surat. When this is applied centrally at the institute, the amount of energy saved will be exceptionally high. This research is a part of the ongoing research and this study is a pilot study. The outstanding outcome obtained by this study will further be an inspiration for other departments and other institutions to conserve energy while saving the money and preserving the environment. The initial cost that is invested to install the smart light are paid back only in a period of 4 to 8 months. Lifetime benefits can be achieved by using the smart technology along with conserving the environment.

Acknowledgement. This research study is a part of Micro Research Project between SVNIT, Surat and MMMUT, Gorakhpur under Twinning Activity of TEQIP-III. We are thankful to TEQIP-III coordinators of SVNIT, Surat and MMMUT, Gorakhpur. We express our heartfelt gratitude to Estate section and also the staff members of CED for providing help and support during field data collection. We also acknowledge the administrative officers of SVNIT, Surat for supporting the research activities and Chatur Lights company for providing and fixing the fixtures.

References

Delaney, D.T., O'Hare, G.M.P., Ruzzelli, A.G.: Evaluation of energy-efficiency in lighting systems using sensor networks. In: Proceedings of the First ACM Workshop on Embedded Sensing Systems for Energy-Efficiency in Buildings (2009)

Janani, R., Kalyana Chakravarthy, P.R., Rathan Raj, R.. A study on value engineering & green building in residential construction. Int. J. Civ. Eng. Technol. **9**(1), 900–907 (2018)

Lockwood, C.: Building the green way. Harv. Bus. Rev. **84**(6), 129–137 (2006)

Wigginton, M., Harris, J.: Intelligent skins. Architectural Press, Oxford (2002)

Magno, M., et al.: A low cost, highly scalable wireless sensor network solution to achieve smart LED light control for green buildings. IEEE Sens. J. **15**(5), 2963–2973 (2014)

Nguyen, T.A., Aiello, M.: Energy intelligent buildings based on user activity: a survey. Energy Build. **56**, 244–257 (2013)

Omar, O.: Intelligent building, definitions, factors and evaluation criteria of selection. Alex. Eng. J. **57**(4), 2903–2910 (2018)

Pandharipande, A., Caicedo, D., Wang, X.: Sensor-driven wireless lighting control: system solutions and services for intelligent buildings. IEEE Sens. J. **14**(12), 4207–4215 (2014)

Thomas, B., Patil, V.: Energy saving approach for buildings. Int. J. Eng. Innov. Technol. **2**(11), 185–191 (2013)

Wong, J.K.W., Li, H., Wang, S.W.: Intelligent building research: a review. Autom. Constr. **14**(1), 143–159 (2005)

Data Mining and Performance Prediction of Flexible Road Pavement Using Fuzzy Logic Theory: A Case of Nigeria

Adekunle Taiwo Olowosulu[1], Jibrin Mohammed Kaura[1],
Abdulfatai Adinoyi Murana[1], and Paul Terkumbur Adeke[2(✉)]

[1] Department of Civil Engineering, Faculty of Engineering, Ahmadu Bello University Zaria,
Zaria, Kaduna State, Nigeria
[2] Department of Civil Engineering, College of Engineering, Federal University of Agriculture
Makurdi, Makurdi, Benue State, Nigeria
adeke.pt@outlook.com

Abstract. The over dependence on road transport system to cater for the fast growing human population in some developing countries like Nigeria has necessitated the need for the development of an efficient and sustainable road pavement management system. This study used data mining techniques namely; Random Forest, Decision Tree and Naive Bayes algorithms to examine the inferred dataset on flexible road pavement performance attributes and surface condition classification in Nigeria. The data mining techniques were used to investigate hidden relationship between pavement performance variables and to authenticate the accuracy of subjective measurements that were used for pavement surface condition classification. The Random Forest and Decision Tree algorithms reported perfect classifications of road pavement sections into; Excellent, Good, Fair, Poor and Very poor. On the other hand, the Naïve Bayes algorithm yielded inaccurate classifications with some margin of errors which were attributed to missing and noisy entries in the dataset. This necessitated the use of Fuzzy logic theory for the performance prediction due to its capability to handle the imprecise dataset. It was used to develop Fuzzy Inference System (FIS) for performance prediction of flexible road pavement using attributes such as; the classified Initial Pavement Condition (IPC), Age of pavement, Resilient Modulus (M_R) of subgrade soil, Average Truck load per day, Average Annual Air Temperature and Rainfall to predict the Future Pavement Condition (FPC). The model was calibrated using the observed logical behaviour of road pavement to fit the engineering experience and judgement. A goodness-of-fit test between the observed and predicted FPC values showed high level of consistency – correlation coefficient at 90%. The research proposed 5120 mutually exclusive Fuzzy Logic Rules for performance prediction of road pavement based on permutation theory. Though, the required well-spread dataset for calibration of the model to cover all possible pavement conditions in Nigeria and subsequent validation were not available, a framework for performance prediction of flexible road pavement was developed, and a comprehensive guidelines on how to calibrate the FIS model using well-spread dataset was presented.

Keywords: Data mining · Random forest · Decision tree · Naive Bayes · Flexible pavement · Fuzzy logic

© The Author(s), under exclusive license to Springer Nature Switzerland AG 2021
H. Shehata and S. El-Badawy (Eds.): *Sustainable Issues in Infrastructure Engineering*, SUCI, pp. 163–192, 2021.
https://doi.org/10.1007/978-3-030-62586-3_11

1 Introduction

The philosophy of pavement usage and preservation states that; it is most cost effective to extend the time when pavements are in good condition with relatively low-cost treatments rather than letting the pavement condition deteriorate until more extensive work is required (Transport Research Board of the National Academies 2007). It is therefore very essential to develop an efficient and sustainable road pavement management system for countries like Nigeria whose major form of transportation is road transport. The practice of pavement management through performance monitoring and prediction for effective planning and timely maintenance and rehabilitation started decades ago. In 1950s, the AASHTO previously known as AASHO first carried out road test program to evaluate pavement conditions in America for efficient maintenance strategy (AASHTO 1993). Thereafter, researches on pavement management system worldwide continued till date (Fwa *et al.* 1994; Fwa and Shanmugam 1998; Cheu *et al.* 2004; Chassiakos 2006; Liu and Sun 2007; Golroo and Tighe 2009; Bianchini and Bandini 2010; Mubaraki 2010; Murana *et al.* 2012a, b; Thube 2011; Mahmood *et al.* 2013; Setyawan *et al.* 2015; Mahmood *et al.* 2015; Savio *et al.* 2016; Ziliute *et al.* 2016; Woo and Yeo 2016; Uglova and Tiraturyan 2016; Xiao and Wu 2016; Premkumar and Vavrik 2016; Pantuso *et al.* 2019; Marcelino *et al.* 2019; Santos *et al.* 2020).

A flexible pavement is an elastic structure made of different unbound layers of soil materials that exhibit nonlinear behaviour under the influence of traffic load, weather condition, material properties, age, etc. A typical cross-section of flexible road pavement structure built according to the Nigeria specifications for road construction is as presented in Fig. 1;

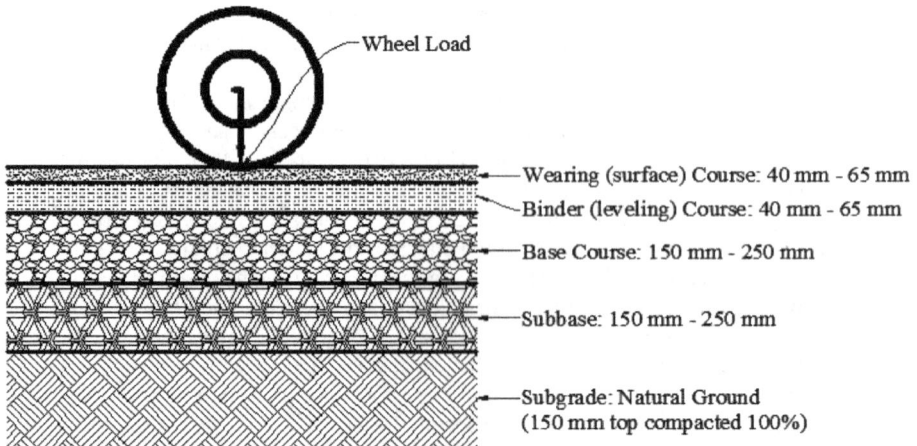

Fig. 1. Schematic of flexible pavement

The choice of flexible pavement for highway construction by most developing countries like Nigeria is due to its relatively cheap cost of construction, maintenance and rehabilitation (Taylor and Philip 2015); though its major disadvantages include relatively short life cycle and rapid deterioration over time (Yin 2007; Ayed 2016; Ziliute *et al.* 2016). The rate of deterioration of flexible pavement depends on several factors ranging from traffic load, climatic condition, material properties, maintenance policy, design of pavement structure and quality of workmanship during construction (Kirbas and Karasahin 2016; Murana 2016; Pantuso *et al.* 2019; Marcelino *et al.* 2019). Development of pavement management system is aimed at achieving good roads that could make cities more liveable and the highways safe for commuting. Threats caused by failed roads include extensive traffic noise and air pollution, massive wear and tear of vehicles, loss of human lives due to fatal auto crashes, high cost of travel due to congestions and fuel consumption, etc. Since huge funds are required for highway construction and maintenance, periodic and timely planning for repairs are essential for efficient management system (Miradi 2009; Abiola *et al.* 2010; Chikezie *et al.* 2011; Owolabi and Abiola 2011; Murana *et al.* 2012a, b; Adefemi and Ibrahim 2015; Mahmood 2015; Ayed 2016; Huang 2017; Gogoi and Dutta 2019).

The assessment of pavement performance usually employs different engineering techniques that require field data generated on-site or obtained from database of pavement management agencies, or from experienced engineers working on pavement distresses (Huang 2004; Miradi 2009; Saltan *et al.* 2011). The performance of road pavement could be classified on a scale of good to worse. The classification is a function of the severity level, quantity and frequency of surface defects caused by deterioration of the pavement materials and its drainage condition (Claros *et al.* 1986; ASTM D6433 2007). The periodic deterioration and consequent failure of road pavement has become a serious concern to pavement management agencies, hence the need for accurate condition classification for optimum performance prediction (Yin 2007; Mahmood 2015; Ayed 2016; Ziliute *et al.* 2016; Adeke *et al.* 2018a, b).

Following the advancement in research activities recently, the use of intelligent algorithms known as Artificial Intelligence (AI) techniques for developing pavement management systems has become a reliable research methodology (Miradi 2009; Russell and Norvig 2010; Mahmood 2015; Marcelino *et al.* 2019; Santos *et al.* 2020). This intelligent approach employs methods of data mining (knowledge discovery) such as Decision Tree, Random Forest, Naïve Bayes classifier, Rough Set Theory, etc. and machine learning (predictions based on known properties) such as; Fuzzy Logic, Genetic Algorithms (GA), Artificial Neural Networks (ANN), etc. (Smadi 2000; Munakata 2008; Witten *et al.* 2020; Fox 2018).

The development of pavement performance prediction models requires input data on pavement surface distress attributes such as; cracks, potholes, rutting, roughness, drainage condition, etc. Other parameters include material strength properties, environmental conditions (climate – rainfall and temperature), traffic load intensity and distribution pattern, and pavement structure are also essential input variables. The process involves generation of huge dataset on pavement behaviour over time (Yu 2005; Setyawan *et al.* 2015; Premkumar and Vavrik 2016; Hamed and Kakarash 2016). Intelligent algorithms are often used for data mining (clustering, classification, visualisation, prediction, etc.) to eliminate challenges of missing or incomplete dataset as relates to data collection and storage for efficient analysis (Claros *et al.* 1986; Miradi 2009; Road Sector Development Team 2014; Luca *et al.* 2016; Fox 2018; Dong *et al.* 2018).

1.1 The Concept of Data Mining

Data mining is a mathematical technique of machine learning which uses black-box models or exploratory techniques to examine dataset that define systems behaviour (Witten and Frank 2005; Fox 2018). It is an approximation technique used for classification, prediction and analysis of imprecise, uncertain or incomplete dataset and knowledge based elements with associated attributes to discover patterns and relationship between variables (Pawlak 2002). Other functions of data mining include to identify partial or total dependencies in a given dataset, eliminates redundant data, gives approach to null values, missing data, dynamic data, etc. The concept is based on the assumption that every object or data point of the universe of discourse is associated with some information (Miradi 2009). It is also believed that objects characterised by the same information are similar in view of their available information. Examples of intelligent algorithms used as classifiers include; Rough Set theory (Miradi 2009), Artificial Neural Network (ANN), Random SubSpace, Support Vector Machines, Pace Regression, Random Forest, Decision Tree (Gopalakrishnan *et al.* 2013; Sharma and Jain 2013), Naïve Bayes theorem (Inkoom *et al.* 2019), etc.

1.1.1 Random Forest

This classifier uses several decision trees in its algorithm to classify systems behaviour during training for accurate classification and predictions. Each tree represents a set of decision alternatives which contest for optimum votes within the dataset (Cigsar and Unal 2019; Li *et al.* 2019). Its works by constructing a group of randomly created decision trees and forecasting the class that is the mode of all classes (classification) or the mean (regression) of the individual trees. The Decision Trees also protect each other from erroneous alternatives since some are weak and unavoidably lead to inaccurate predictions. A collection of weak decision trees can yield better predictions since the prediction accuracy depends on the number of decision trees and predators used in the model. Random Forest has been used by previous researchers for interpretation, visualization, and handling complex non-linearity behaviour between predictors and response. Its limitations include over-fitting, not robust to outliers, and poor in predictive capacity compared to methods like the least square regression. The major advantage of the Random Forest algorithm is that, it does not permit overfitting (Kudjo *et al.* 2020). Li *et al.* 2019 used Random Forest algorithm for the identification of asphalt pavement distresses and condition classification to aid pavement management practice. Findings of the study confirmed its suitability due to relatively high degree of accuracy obtained as compared to the maximum likelihood classification and the support vector maching models. Gong *et al.* (2018) revealed that Random Forest algorithm significantly outperformed the linear regression model with coefficients of determination greater than 0.95 in both training and test sets when used for predicting the IRI of pavement. Pan *et al.* (2018), also used the Random Forest algorithm to detect potholes and caracks in asphlt pavement, and it performance was commendable.

1.1.2 Decision Trees

This is a flow-chart like structure with internal nodes which denote tests on variable sets, each branch represents an outcome of the test and the leaf node represents class distribution. The trees are generated from training data in a top-down, general-to-specific direction. The mechanism of decision tree is such that, whenever all data points have the same output or belong to the same class, there is no further decisions to be made with respect to partition of the data points, and the solution is complete. Contrary to when data points at a node have different outputs or belong to two or more classes, then a test is made at the node that will result into a split. The process is recursively repeated for each of the new intermediate nodes until a completely discriminating tree is obtained. The process is potentially an over-fitting solution, hence eliminates components that are too specific to noise and outliers that may be present in the training data (Miradi 2009; Gopalakrishnan *et al.* 2013).

The recent advanced model used for Decision Tree is kown as the J48 algorithm. Its structure is made of root or internal node which represents array of decision attributes, branches of the tree represent observations of the system performance based on possible outcomes and the leaves or terminal nodes are the target values (regression) or final class (classification) of the dependent variable. Estimations of J48 involve the manipulation of missing values, pruning, derivation of rules, etc. It uses predictive machine-learning model to calculate the resultant value of a new sample based on various attribute values of the available data. A decision tree can handle both binary and non binary classification (Ahishakiye *et al.* 2017; Cigsar and Unal 2019; Obuandike *et al.* 2015). Saravanan and Gayathri (2018) identified limitations of J48 algorithm to include; missing values of zero entries which make the tree wider and complicated, irrelevant entries to a class which lead to wrong classification and the presence of noisy data element which cause over fitting thereby increasing the margin of error.

On the other hand, according to Miradi (2009), advantages of the Decision Tree algorithm include; to handle both numerical and categorical variables, carryout classification/regression in simple form, stores old dataset and efficiently manipulated new dataset, carryout automatic stepwise variables selection and complexity reduction, extremely robust in handling outliers and easy to understand and interpret results. Cubero-Fernandez *et al.* (2017) employed the use of Decision Tree algorithms for pavement crack detection and classification into transverse longitudinal and alligator cracks. The study recommend the decision tree algorithm for further analysis due to its high level of accuracy. Also, Lin *et al.* (2013), Gopalakrishnan *et al.* (2013) and Inkoom *et al.* (2018) used decision tree algorithms for developing pavement maintenance and management system, results obtained by the algorithms yielded accurate estimations.

1.1.3 Naive Bayes Theorem

This algorithm is based on the Bayesian probability theorem which operates on conditional rules (Marianingsih and Utaminingrum, 2018; Inkoom *et al.* 2019). It is an intuitive approach which uses conditional probabilities of the occurrence of individual attributes in a given class label for making prediction of the targeted attribute. It could be used for attribute classifications, the Naïve Bayes algorithm is mostly used for predictive modelling using discrete events of a database without missing data to predict class labels (Alam and Pachauri 2017). It is an example of the Bayesian classification which computes or predictes the probability of an attribute belonging to a given class based on the assumption that the effect of that attribute is independent of the other attributes in the same class, hence is known as class conditional independence model (Tribhuvan *et al.* 2015). According to Alam and Pachauri (2017) the Bayes theorem calculates the probability of a targeted attribute by counting the frequency and combinations of values in the dataset, then estimating the parameter using method of maximum likelihood. The algorithm suites most complex real world problems. Also, its ability to estimate the parameter using small amount of training data is considered the major advantages of the algorithm. But its robustness is easily affected by noisy element in the dataset. According to Jang *et al.* (2015), the Naïve Bayes model is expressed as shown in Eq. 1;

$$P(y|x) = \frac{P(x|y).P(y)}{P(x)} \tag{1}$$

where $P(y|x)$ is the posterior probability of class y (target) given the predictor x (attribute), $P(x)$ is the prior probability of predictor, $P(y)$ is the prior probability of class and $P(x|y)$ is the likelihood which is the probability of predictor given class. Previous studies that used the Naïve Bayes algorithm in pavement management system recorded significant successes (Marianingsih and Utaminingrum 2018; Marianingsih *et al.* 2019; Inkoom *et al.* 2019); its inability to handle missing data and noisy data points yielded results with relatively significant margin of error (Shekharan 1998; Jang *et al.* 2015; Alam and Pachauri 2017; Gong *et al.* 2018).

1.2   The Concept of Fuzzy Logic Theory

Fuzzy Logic is an AI technique used for developing knowledge-based models which use human intuitive reasoning characterised by subjectivity and imprecision. Unlike the classic logic defined by binary values of 0 and 1, it is a multivalued logic model capable of measuring intermediate values between conventional evaluations of *on or off, true or false, yes or no,* etc. which could be formulated mathematically and processed using computers (Munakata 2008). Fuzzy Logic describes a system at varying degrees using adjectives like *'low', 'medium' and 'high'*. The theory itself is not imprecise in nature, but a mathematical theory which deals with subjectivity and uncertainties. It is capable of handling data inaccuracies, uncertainties and non-linearity. It uses intuitive and subjective expert knowledge or experience for Fuzzification process where there are no sufficient numeric data. The Fuzzy rules try to execute Fuzzy reasoning used for evaluating true degree of goal propositions (Hamed and Kakarash 2016). In this theory, the human vague thought and perception on interpretation of systems behaviour is expressed using appropriate mathematical models. Although probility theorems such as stochastic or Monte Carlo theorems have vast applications in analysing varying systems, they are subsets of the Fuzzy theorem (Bai *et al.* 2006; Munakata 2008). Fuzzy logic provides mathematical strength for emulating certain perceptual and linguistic attributes associated with human reasoning which are a function of experience over time in a systematic manner. Some quantities do not have sharp boundaries, hence best described using labels as; *'worse'*, *'many'*, *'tall'*, *'young'*, *'small'*, etc. which represent the fact known as Fuzzy concepts. They are usually true to some degree that cannot be easily quantified or are relative in nature as well as false on the other hand. The major limitation of Fuzzy logic is caused by uncertainties associated with total lack of information about the system behaviour. In contrast, the traditional binary logic usually has discrete set of options used for describing crisp events without intermediate classes.

1.2.1 Fuzzy Logic Modelling

A Fuzzy inference-based system is an intelligent method used for classification and prediction problems. The system interprets values in the input vector based on user-defined rules and assigns values to the output vector. The advantages of this approach is its knowledge representation in form of the *IF-THEN* logic rules, the mechanism of reasoning in human understandable terms, the capacity of taking linguistic information from human experts and combining it with numerical information, and the ability of approximating complicated nonlinear functions with simpler models. The Fuzzy rules are usually generated either from an expert knowledge or numerical dataset which makes the theory suitable for analysing several problems (Munakata 2008; Mahmood *et al.* 2013). A Fuzzy set is an extension of a crisp set since crisp set allows only full membership or no membership at all, a Fuzzy set allows for partial membership in the system (Bai *et al.* 2006). The Fuzzy Inference System (FIS) depends on its Membership Function (MF) which defines its correct value between 0 and 1. The degree to which any Fuzzy statement is true falls within the range of 0 and 1. Fuzzy logic provides an inference mechanism under cognitive uncertainty and mathematical computations using words. According to Chen and Pham (2001), for a conventional set, the MF of an element x in subset A, $\mu_A(x)$ is expressed as Eq. 2;

$$\mu_A(x) = \begin{cases} 1 & \text{if and only if } x \in A \\ 0 & \text{otherwise} \end{cases} \tag{2}$$

The degree of membership of each entity is a numerical measure from 0 to 1, where 0 denotes nonexistence of the element in the subset or set, 1 denotes full membership of the set and any value between 0 and 1 defines the partial degree of membership of the element in the set (Chen and Pham 2001; Bai *et al.* 2006). The basic mathematical relationship between subsets A and B include; union, intersect and complements, which are mathematically expressions as Eqs. 3–5;

$$A \cup B = \mu_A(x) \cup \mu_B(x) = \max(\mu_A(x), \mu_B(x)) \tag{3}$$

$$A \cap B = \mu_A(x) \cap \mu_B(x) = \min(\mu_A(x), \mu_B(x)) \tag{4}$$

$$A^c = 1 - \mu_A(x) \tag{5}$$

The Fuzzy MFs are commonly expressed graphically as; triangular, trapezoidal, Gaussian, bell-shaped, sigmoid, etc. curves. According to Mahmood *et al.* (2015), Santos *et al.* (2020) and Gogoi and Dutta (2019), the triangular MFs is suitable for pavement performance analysis. Chen and Pham (2001) and Bai *et al.* (2006) defined mathematical details of the triangular MF as shown in Fig. 2;

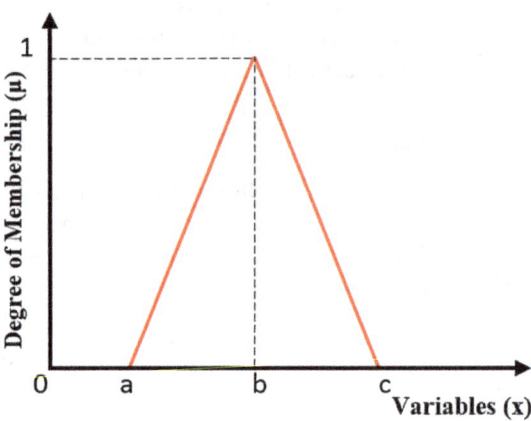

Fig. 2. Triangular Fuzzy MF

Figure 2 presents a triangular MF of Fuzzy set A, where only elements at point b assume full membership of 1. Models that define the degree of membership of elements of Fuzzy set A at various stages are as shown in Eqs. 6a and 6b;

$$\mu_A(x) = \begin{cases} 0 & x \leq a \\ \frac{x-a}{b-a} & a \leq x \leq b \\ \frac{c-x}{c-b} & b \leq x \leq c \\ 0 & x \geq c \end{cases} \tag{6a}$$

$$\mu_A(x; a, b, c) = max\left(min\left(\frac{x-a}{b-a}, \frac{c-x}{c-b}\right), 0 \right) \tag{6b}$$

The Mamdani framework is commonly used for executing FIS based on its set of MFs for both input and output variables in form of crisp values (Kaur and Tekkedil 2000; MATLAB 2015).

1.3 Analysis of Systems Behaviour Using Fuzzy Logic

Following guidelines stated by Bai *et al.* (2006), the three basic steps required for solving real life problems using the Fuzzy logic approach include Fuzzification, formation of Fuzzy logic rules and Defuzzification. Fuzzification of quantities is the process of translating the available numerical input variables into linguistic variable such as low, medium and high based on defined thresholds of membership or classifications of Fuzzy sets used for the development of Fuzzy logic rules. The Fuzzy logic rules are the decision rules based on experience and logic applied on the Fuzzified linguistic variables. Defuzzification of quantities is the end product which reverses the Fuzzification process to produce actual and meaningful output in linguistic form. The process involves reconverting the linguistic terms into crisp values according to their degree of membership using Fuzzy logic model such as the Mamdani model. According to Chen and Pham (2001) and Bai *et al.* (2006) some numerical methods of Defuzzification analysis include; the centre of gravity method, mean of maximum method, height method, etc. The centre of gravity method is commonly used, it defines the defuzzified value $\tilde{x}$ of a Fuzzy set as the Fuzzy centroid measured along the horizontal axis expressed as the centre of area under the MF curve using Eqs. 7a and 7b for discrete and continuous systems analysis respectively;

$$\tilde{x} = \frac{\sum_{i=1}^{n} x_i . \mu_i(x_i)}{\sum_{i=1}^{n} \mu_i(x_i)} \tag{7a}$$

$$\tilde{x} = \frac{\int_1^n x_i . \mu_i(x_i) dx}{\int_1^n \mu_i(x_i) dx} \tag{7b}$$

where, $\mu_i(x_i)$ is the MFs of area of the i th sampled element in the Fuzzy set and x_i is its centroid distance from the origin of the horizontal axis or crisp value.

1.4 Pavement Management System Using Fuzzy Logic

The subjective measurement of pavement performance attributes for other prediction models is seen as major limitation since its creates significant margin of error in estimation, this disadvantage is better handled using the Fuzzy logic theory due to its ability to Fuzzify variables (Liu and Sun 2007; Karagahin and Terzi 2014; Mahmood *et al.* 2015). Some existing pavement performance prediction models are based on pavement surface condition classification such as; Condition Rating Survey (CRS) and Pavement Condition Index (PCI) methodologies among others (Dabous *et al.* 2019a, b). The use of such models require the initial subjective assessment of pavement distress conditions as input variable (Setyawan *et al.* 2015; Premkumar and Vavrik 2016; Hamed and Kakarash 2016). In Nigeria, dataset on road pavement surface condition evaluation are not readily available. This is basically due to the cost of operation, lack of machineries and operational skills, government policies, etc. Though the challenge is not peculiar to Nigeria (Dabous *et al.* 2019a, b). Investigating the impact of various damaging factors on road pavement require large data on all the influencing factors which is usually expensive in terms of cost and time (Abiola 2012; Surendrakuma *et al.* 2013; Mane *et al.* 2016). Salpisoth (2014) identified major challenges in road pavement management system to include; the ability to carry out routine monitoring and evaluation of pavement condition regularly with limited budget and expertise; and how to predict pavement performance using appropriate models based on incomplete data for optimum maintenance and repair decisions. The use of Fuzzy logic method in pavement management system in terms of condition classification, performance predictions and decision prioritisation has gained wide commendation in recent times. This assertion can be proven using previous studies as listed in Table 1;

This study therefore aims at developing a framework for data mining and performance prediction of flexible road pavement using Fuzzy Logic theory. Objectives of the

Table 1. The use of fuzzy logic theory in pavement management system

Author(s) and date	Methodology	Findings
Shoukry *et al.* (1997)	Pavement condition classification	Recommended
Fwa and Shanmugam (1998)	Pavement condition classification	Recommended
Bandara and Gunaratne (2001)	Pavement condition classification	Recommended
Arliansyah *et al.* (2003)	Pavement condition classification	Recommended
Golroo and Tighe (2009)	Pavement condition classification	Recommended
Koduru *et al.* (2010)	Pavement condition classification	Recommended
Mahmood *et al.* (2013)	Pavement condition classification	Recommended
Mahmood (2015)	Pavement condition classification	Recommended
Mahmood *et al.* (2015)	Pavement condition classification	Recommended
Kaur and Tekkedil (2000)	Pavement performance prediction	Recommended
Liu and Sun (2007)	Pavement performance prediction	Recommended
Miradi (2009)	Pavement performance prediction	Recommended
Bianchini and Bandini (2010)	Pavement performance prediction	Recommended
Wang and Li (2011)	Pavement performance prediction	Recommended
Karagahin and Terzi (2014)	Pavement performance prediction	Recommended
Aggarwal and Kumar (2015)	Pavement performance prediction	Recommended
Jeong *et al.* (2017)	Pavement performance prediction	Recommended
Cheu *et al.* (2004)	Maintenance decision prioritisation	Recommended
Chassiakos (2006)	Maintenance decision prioritisation	Recommended
Chandran et al. (2007)	Maintenance decision prioritisation	Recommended
Chen and Flintsch (2008)	Maintenance decision prioritisation	Recommended
Moazami *et al.* (2011)	Maintenance decision prioritisation	Recommended
Gogoi and Dutta (2019)	Maintenance decision prioritisation	Recommended
Nawir and Prihartanto (2019)	Maintenance decision prioritisation	Recommended
Santos *et al.* (2020)	Maintenance decision prioritisation	Recommended

study include; to carryout data mining for optimum initial road pavement surface condition classification based on attributes that affect road pavement performance in Nigeria using intelligent algorithms of Waikato Environment for Knowledge Analysis (WEKA) software, to develop and simulate a Fuzzy Logic Inference System for performance prediction of flexible road pavement in Nigeria implemented using MATLAB software and to develop a decision framework for performance prediction of flexible road pavement in Nigeria.

2 Methodology

2.1 Description of Study Area

Nigeria is among the most populous countries on the Africa continent with an estimated human population of 206 million persons. Its land mass is estimated at 923,768.00 square kilometres of surface area. Road transport is her major form of land transportation system. The distribution of Federal Highways in Nigeria by length sums to 34,340.95 km (Federal Government of Nigeria 2012). Details of some selected links of Federal Highway in Nigeria is as shown in Table 2;

Table 2. Selected federal highway links in Nigeria (Claros *et al.* 1986)

S/N	State	Link ID	Length (km)	No. of lanes	Age (years)	Surface finishing
1.	Anambra	89	43.7	2	6	Asphalt concrete
2.	Bauchi	285	51.3	2	8	Asphalt concrete
3.	Benue	112	53.6	2	6	Asphalt concrete
4.	Borno	716	50.7	2	9	Surfaced dressed
5.	Kaduna	130	46.8	2	5	Asphalt concrete
6.	Kwara	5	22.8	2	9	Asphalt concrete
7.	Ogun	17	18.3	2	12	Asphalt concrete
8.	Ogun	332	30.9	2	10	Asphalt concrete
9.	Plateau	136	10.0	2	9	Asphalt concrete
10.	Plateau	138	23.2	2	5	Asphalt concrete
11.	Imo	370	45.6	2	2	Asphalt concrete
12.	Rivers	144	18.0	2	8	Asphalt concrete
13.	Sokoto	255	57.3	2	8	Asphalt concrete
14.	Oyo	22	23.4	2	10	Asphalt concrete

2.2 Data Collection and Description

The study considered secondary dataset obtained from the database of pavement condition evaluation unit of the Federal Ministry of Power, Works and Housing Nigeria as reported in Claros *et al.* (1986) for data mining and model development. The dataset captured condition attributes of 102 links of flexible pavement of Federal Highways across Nigeria. The relevant attributes used for the study included;

1. **Average daily truck load:** this was reported as the average number of daily truck load known as heavy goods or commercial vehicles in each direction that weighed at least 1500 kg as estimated from the Average Daily Traffic (ADT) volume travelling on the Federal Highway facility during the investigation period.

2. **Strength property of subgrade soil:** the Resilient Modulus (M_R) of subgrade soil determined based on procedures stated in AASHO T-274-82 as reported by Claros *et al.* (1986) was used as the subsoil strength representative variable.
3. **Weather condition:** weather parameters of interest included the Average Annual Air Temperature and Rainfall within the classified Zones.
4. **Age of pavement:** this was the period in years measured from when the Highway facility was first opened to traffic or rehabilitated till the date of condition evaluation survey.
5. **Initial pavement condition (IPC):** this classification used the Condition Survey Rating Scores (CSRS) which measures from 0–100 for Very poor to Excellent based on the average rut depth (cm), percentage area of alligator or fatigue cracking and the drainage condition per road segment. The classifications of CSRS scale is as presented in Table 3;

Table 3. Classification of pavement surface condition

Classification of pavement condition	Limits of CSRS
Excellent	>81
Good	66–81
Fair	46–65
Poor	25–45
Very poor	<25

2.3 Data Mining for IPC Classification

Data mining algorithms were used to determine the accuracy of initial surface condition classification of pavement using WEKA software since the measure of some attributes was subjective and characterised by missing and incomplete dataset. The Random Forest, Decision Tree and Naïve Bayes classifiers were used for optimum training (Witten *et al.* 2015). Aside the dataset used for IPC classification which was vague in nature, other performance attributes such as; traffic load, temperature and rainfall, etc. had no sufficient dataset that necessitated data mining process, hence were manually classified based on guidelines specified in manuals and experience (Claros *et al.* 1986; Road Sector Development Team 2014). Table 4 presents the summary of data attributes and types used (Claros *et al.* 1986);

Table 4. Data attributes and type

S/N	Attributes	Data type
1.	Average rut depth (cm)	Numerical
2.	Fatigue cracking (%)	Numerical
3.	Drainage condition	Nominal
4.	CSRS (%)	Numerical
5.	Pavement condition	Nominal

2.4 Development of FIS

The Fuzzy logic tool in MATLAB programming software was used to develop Mamdani FIS. The Fuzzification process involved identifying essential attributes into Fuzzy sets as input and output for performance prediction of road pavement. The creation of Fuzzy MF involved identifying suitable function (curve) based on classified attributes in the datasets, then calibrating it using representative parameters of the attribute (Mahmood 2015; Nawir and Prihartanto 2019; Santo *et al.* 2020). Table 5 presents the proposed classifications of attributes and ranges to form Fuzzy sets used for the development of Fuzzy MF;

Table 5. Development of fuzzy sets for attributes

S/N	Parameters	Classifications	Limits	Ranges of FMF
1.	Initial pavement condition (IPC)	Very poor	<25	[0 15 25]
		Poor	25–45	[20 35 50]
		Fair	46–65	[40 50 65]
		Good	66–81	[60 70 87]
		Excellent	>81	[80 90 100]
2.	Age of pavement (years)	New	<2	[0 1 3]
		Recent	2–5	[2 5 7]
		Old	6–12	[6 9 13]
		Very old	>12	[12 16 20]
3.	Truck load (veh/day)	Low	<200	[0 125 280]
		Medium	200–1000	[195 600 1000]
		High	1001–2000	[850 1400 2000]
		Very high	>2000	[1700 2000 2500]
4.	M_R of subgrade soil (kg/m^2)	Low	$<0.7 \times 10^7$	$[0\ 0.4\ 0.75] \times 10^7$
		Medium	$[0.7–1.3] \times 10^7$	$[0.6\ 1.0\ 1.425] \times 10^7$
		High	$[1.3–2.1] \times 10^7$	$[1.2\ 1.8\ 2.4] \times 10^7$
		Very high	$>2.1 \times 10^7$	$[2.0\ 2.48\ 3.0] \times 10^7$
5.	Temperature ($^\circ$C)	Low	<20	[0 10 20]
		Medium	20–35	[15 25 35]
		High	36–45	[30 35 42]
		Very high	>45	[40 45 50]
6.	Rainfall (mm)	Low	<600	[0 250 650]
		Medium	600–1200	[500 820 1200]
		High	1201–1800	[1100 1500 1900]
		Very high	>1800	[1700 2000 2500]
7.	Future pavement condition (FPC)	Excellent	>81	[0 15 25]
		Good	66–81	[20 35 50]
		Fair	46–65	[40 50 65]
		Poor	26–45	[60 70 87]
		Very poor	<25	[80 90 100]

Using Fuzzy sets presented in Table 5, the Fuzzification and creation of Fuzzy logic rules for implementation using MATLAB software based on the data source was as shown in Table 6;

Table 6 revealed that highway links whose properties satisfied a given trend or had similar pattern of logical behaviour were identified and grouped into blocks with the observed crisp values that defined the Future Pavement Condition (FPC). At the instances of missing data and pattern mismatch, simple extrapolation techniques (guided by expe-

Table 6. Fuzzification and fuzzy logic rules

Rule No.	Link Code	IPC	Age (Years)	Truck Load (Veh/Day)	MR of Subgrade (kg/m2)	Temp. (°C)	Rainfall (mm)	FPC	Crisp Value
1.	370	Excellent	New	Low	High	Medium	V.High	Good	76
2.	615	Excellent	Recent	Low	High	Medium	High	Excellent	85
3.	61	Excellent	Recent	Medium	Medium	Medium	High	Excellent	93
4.	565	Excellent	Recent	Medium	Medium	Medium	V.High	Excellent	87
5.	130	Excellent	Recent	Medium	High	Medium	High	Excellent	92
6.	298	Excellent	Recent	Medium	High	Medium	V.High	Good	72
7.	736	Excellent	Recent	Medium	V.High	Medium	High	Good	78
8.	405	Excellent	Old	Low	Low	Medium	High	Good	76
9.	371	Excellent	Old	Low	Medium	Medium	V.High	Good	76
10.	716	Excellent	Old	Low	Medium	Medium	Medium	Fair	60
11.	27	Excellent	Old	Low	Medium	Medium	Medium	Excellent	92
12.	472	Excellent	Old	Low	High	Medium	Medium	Excellent	86
13.	287	Excellent	Old	Low	High	Medium	Medium	Good	77
14.	551	Excellent	Old	Low	High	Medium	High	Excellent	90
15.	630	Excellent	Old	Low	High	Medium	V.High	Good	78
16.	289	Excellent	Old	Medium	Medium	Medium	Medium	Excellent	83
17.	219	Excellent	Old	Medium	Medium	Medium	High	Excellent	94
18.	136	Excellent	Old	Medium	Medium	Medium	High	Good	75
19.	252	Excellent	Old	Medium	High	Medium	Medium	Excellent	86
20.	290	Excellent	Old	Medium	High	Medium	Medium	Good	81
21.	17	Excellent	Old	Medium	High	Medium	High	Good	69
22.	275	Excellent	Old	Medium	High	Medium	High	Excellent	90
23.	176	Excellent	Old	High	Medium	Medium	High	Excellent	92
24.	50	Excellent	Old	Medium	High	Medium	V.High	Excellent	92
25.	186	Excellent	Old	Medium	High	Medium	V.High	Good	76
26.	211	Excellent	Old	High	High	Medium	High	Excellent	84
27.	144	Excellent	Old	High	High	Medium	V.High	Good	76
28.	187	Excellent	Old	High	V.High	Medium	V.High	Good	76
29.	173	Excellent	Old	V.High	High	Medium	High	Good	78
30.	210	Excellent	Old	V.High	High	Medium	V.High	Good	79
31.	121	Excellent	Old	High	High	Medium	V.High	Excellent	85
32.	89	Excellent	Old	High	High	Medium	High	Excellent	88
33.	91	Excellent	Old	High	Medium	Medium	High	Excellent	86
34.	123	Excellent	V.Old	Low	High	Medium	V.High	Fair	59

rience) were used to fill some targeted classes that were necessary to aid logical decisions (Chen and Flintsch 2008). The process filtered identical rules and eliminated repeated entries and entries with missing data leading to the developing of 34 unique Fuzzy logic rules (out of 102 links considered) for implementation using MATLAB software. A plot of triangular Fuzzy logic membership functions MF for inputs and output variables used is as shown in Fig. 3;

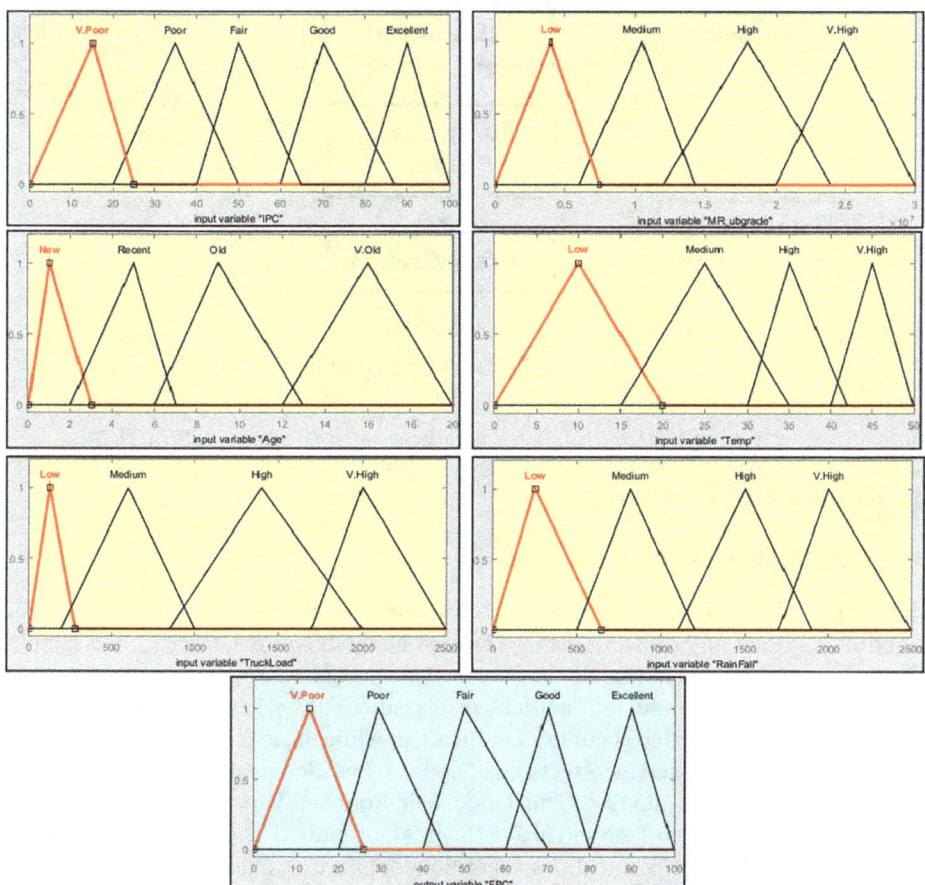

Fig. 3. MF plots for inputs and output variables

The *IF . . . THEN* . . . logic rule was further used to build conditional logic that defined pavement behaviour over time, then the model was simulated and defuzzified to yield FPC as the output. An illustration of this methodology is as presented in Fig. 4;

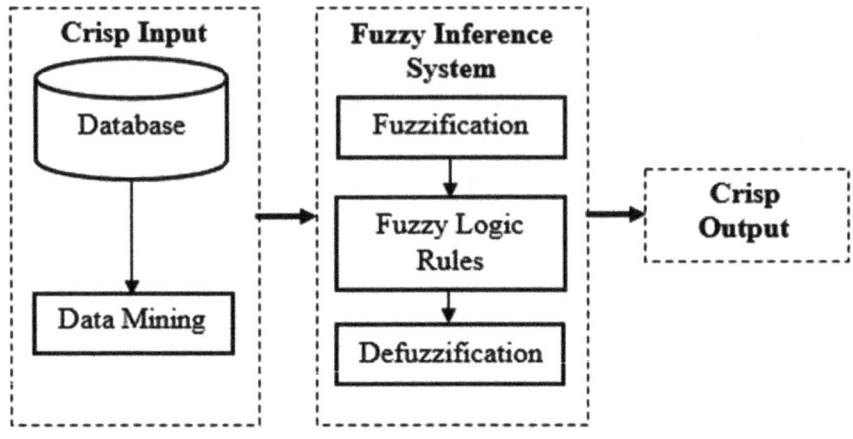

Fig. 4. Schematic of methodology

3   Results and Discussion

3.1   Data Mining for IPC Classifications

Detailed results of initial pavement surface condition classification using intelligent algorithms - Random Forest, Decision Tree and Naïve Bayes classifier for the sampled links were as presented in Fig. 5;

Figure 5 present confusion matrices of classifications where entries with perfect diagonal matrices indicated accurate classification, while those forming upper and lower triangular matrices showed incorrect classifications. Instances considered were correctly or incorrectly classified into Excellent, Good, Fair, Poor and Very Poor conditions by the Random Forest, Decision Tree and Naïve Bayes algorithms. The perfect diagonal matrices for Fig. 5 (a) to (n) classifications using Random Forest model showed its weakness and insensitivity in handling missing and noisy data elements, while the Decision Tree model failed to attain perfect diagonal matrices for classifications of Fig. 5 (a), (h), (m) and (n) to show its relatively low sensitivity to discrepancies in the dataset. On the other hand, the Naïve Bayes model had perfect diagonal matrix at Fig. 5 (j) classification only. This implied that all instances were fittingly classified into the defined classifications of the pavement surface condition based on the attributes used. The Random Forest and Decision Tree classifiers present relatively perfect diagonal matrices while the Naïve Bayes classifier gave substantially dispersed matrices. The dispersed trend of confusion matrices is an indication of incorrect classifications, and is attributed to high sensitivity of the Naïve Bayes classifier to the presence of missing data and noisy data elements in the dataset, which affected the accuracy of its classifications (Inkoom *et al.* 2019; Kudjo

Random Forest	Decision Tree	Naive Bayes

```
=== Confusion Matrix ===

  a   b   c   <-- classified as
 12   0   0 |  a = Good
  0  93   0 |  b = Excellent
  0   0   1 |  c = Fair
```

```
=== Confusion Matrix ===

  a   b   c   <-- classified as
 12   0   0 |  a = Good
  0  93   0 |  b = Excellent
  1   0   0 |  c = Fair
```

```
=== Confusion Matrix ===

  a   b   c   <-- classified as
 10   2   0 |  a = Good
  6  87   0 |  b = Excellent
  0   0   1 |  c = Fair
```

(a) Link 89 in Anambra State

```
=== Confusion Matrix ===

  a   b   c   d   e   <-- classified as
 51   0   0   0   0 |  a = Fair
  0  35   0   0   0 |  b = Poor
  0   0  35   0   0 |  c = Excellent
  0   0   0  42   0 |  d = Good
  0   0   0   0  10 |  e = Very Poor
```

```
=== Confusion Matrix ===

  a   b   c   d   e   <-- classified as
 51   0   0   0   0 |  a = Fair
  0  35   0   0   0 |  b = Poor
  0   0  35   0   0 |  c = Excellent
  0   0   0  42   0 |  d = Good
  0   0   0   0  10 |  e = Very Poor
```

```
=== Confusion Matrix ===

  a   b   c   d   e   <-- classified as
 43   2   0   6   0 |  a = Fair
  2  27   0   0   6 |  b = Poor
  0   0  34   1   0 |  c = Excellent
  0   0   1  41   0 |  d = Good
  0   1   0   0   9 |  e = Very Poor
```

(b) Link 716 in Borno State

```
=== Confusion Matrix ===

  a   b   <-- classified as
 69   0 |  a = Excellent
  0   5 |  b = Good
```

```
=== Confusion Matrix ===

  a   b   <-- classified as
 69   0 |  a = Excellent
  0   5 |  b = Good
```

```
=== Confusion Matrix ===

  a   b   <-- classified as
 68   1 |  a = Excellent
  0   5 |  b = Good
```

(c) Link 5 in Kwara State

```
=== Confusion Matrix ===

  a   b   c   <-- classified as
 21   0   0 |  a = Excellent
  0  16   0 |  b = Good
  0   0  10 |  c = Fair
```

```
=== Confusion Matrix ===

  a   b   c   <-- classified as
 21   0   0 |  a = Excellent
  0  16   0 |  b = Good
  0   0  10 |  c = Fair
```

```
=== Confusion Matrix ===

  a   b   c   <-- classified as
 21   0   0 |  a = Excellent
  0  16   0 |  b = Good
  0   2   8 |  c = Fair
```

(d) Link 370 in Imo State

```
=== Confusion Matrix ===

  a   b   <-- classified as
  6   0 |  a = Good
  0  99 |  b = Excellent
```

```
=== Confusion Matrix ===

  a   b   <-- classified as
  6   0 |  a = Good
  0  99 |  b = Excellent
```

```
=== Confusion Matrix ===

  a   b   <-- classified as
  6   0 |  a = Good
  1  98 |  b = Excellent
```

(e) Link 332 in Ogun State

```
=== Confusion Matrix ===

  a   b   <-- classified as
  7   0 |  a = Good
  0  70 |  b = Excellent
```

```
=== Confusion Matrix ===

  a   b   <-- classified as
  7   0 |  a = Good
  0  70 |  b = Excellent
```

```
=== Confusion Matrix ===

  a   b   <-- classified as
  7   0 |  a = Good
  2  68 |  b = Excellent
```

(f) Link 285 in Bauchi State

```
=== Confusion Matrix ===

   a    b   <-- classified as
 186    0 |  a = Excellent
   0   14 |  b = Good
```

```
=== Confusion Matrix ===

   a    b   <-- classified as
 186    0 |  a = Excellent
   0   14 |  b = Good
```

```
=== Confusion Matrix ===

   a    b   <-- classified as
 184    2 |  a = Excellent
   1   13 |  b = Good
```

(g) Link 255 in Sokoto State

```
=== Confusion Matrix ===

  a   b   c   <-- classified as
 29   0   0 |  a = Good
  0  26   0 |  b = Excellent
  0   0   1 |  c = Fair
```

```
=== Confusion Matrix ===

  a   b   c   <-- classified as
 29   0   0 |  a = Good
  0  26   0 |  b = Excellent
  1   0   0 |  c = Fair
```

```
=== Confusion Matrix ===

  a   b   c   <-- classified as
 27   2   0 |  a = Good
  1  25   0 |  b = Excellent
  0   0   1 |  c = Fair
```

(h) Link 22 in Oyo State

Fig. 5. Surface condition classification of pavement using confusion matrices

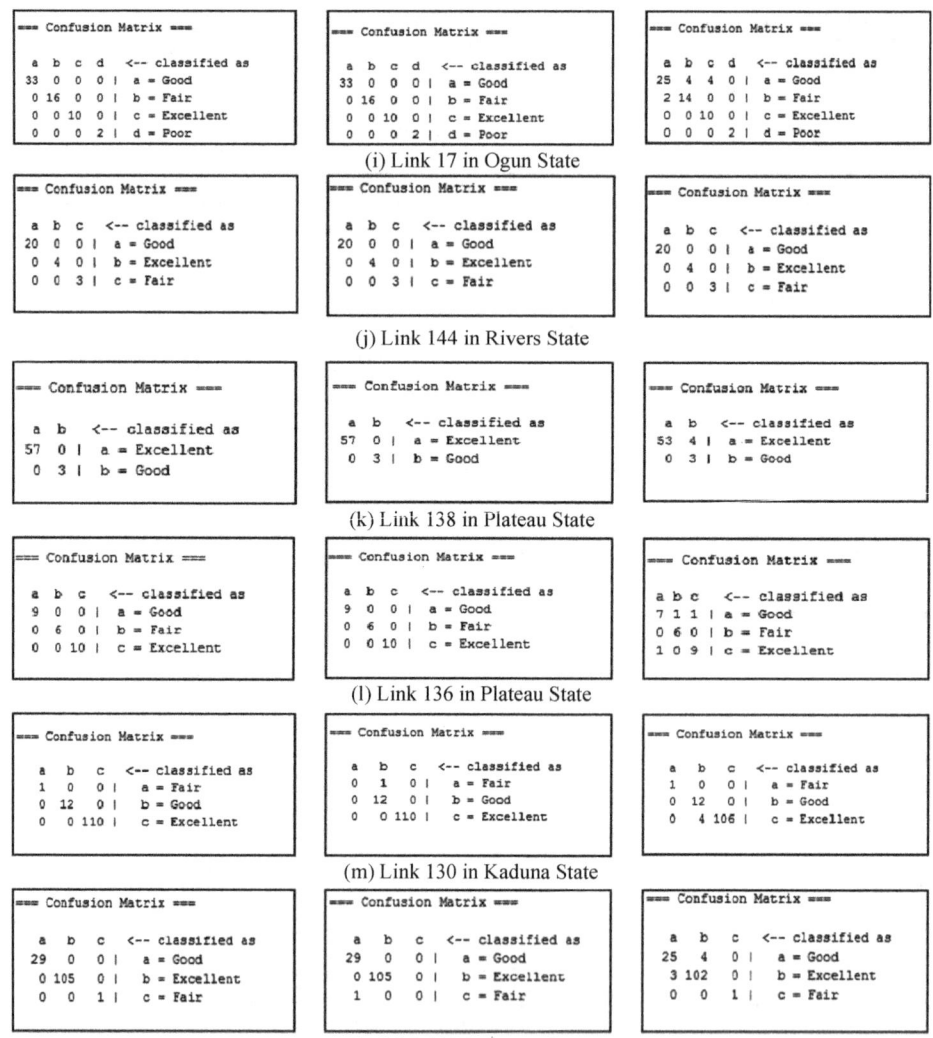

(i) Link 17 in Ogun State

(j) Link 144 in Rivers State

(k) Link 138 in Plateau State

(l) Link 136 in Plateau State

(m) Link 130 in Kaduna State

(n) Link 112 in Benue State

Fig. 5. (*continued*)

et al. 2020). In order words, failure to achieve perfect classification was attributed to the number of incorrectly classified instances due to significant disparities observed in the dataset which were possibly due to missing or noisy data entries (Witten *et al.* 2016; Marianingsih and Utaminingrum 2018). Models performance through percentage accuracy of classifiers was estimated using the correct and incorrect classifications as shown in Fig. 6;

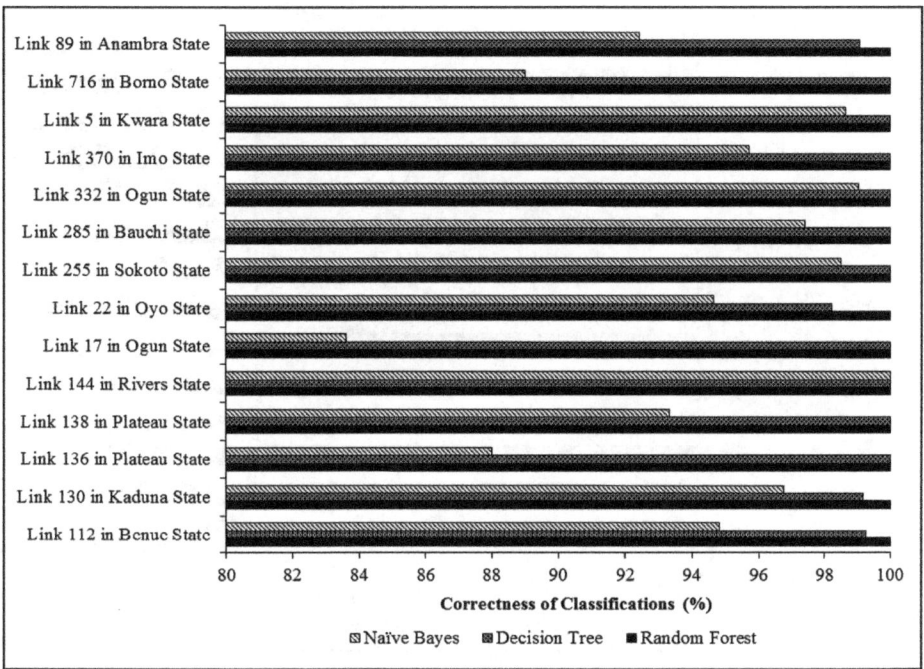

Fig. 6. Accuracy of classifications

Figure 6 revealed that the Random Forest and Decision Tree models made classifications with relatively high accuracy, while the Naïve Bayes classifier made relatively
significant number of incorrect classifications. This was attributed to their insensitivity
to the challenge of missing data compared to the Naïve Bayes algorithms which recorded
percentage errors due to incorrect classification caused by noise data points except for
Link 144 in River State (Shekharan 1998; Jang *et al*. 2015; Alam and Pachauri 2017;
Gong *et al*. 2018; Inkoom *et al*. 2019). Also, the coefficient of correlations analysis of
classifications presented by Kappa statistics of WEKA software is as shown in Fig. 7;

Figure 7 revealed that classifications by the Random Forest model assumed accurate
performance at 100% while those by the Decision tree model had some margin of error
with that of Naïve Bayes theorem having relatively low accuracy, except for link 144
in River state. The classification errors are further quantified using the Root Relative
Squared Error (%) as shown in Fig. 8;

Results presented in Fig. 8 revealed that assertions established in Fig. 7 were correct
due to the inversely proportional relationship between the coefficient of correction and
the estimated root relative squared errors (%) of classification outputs of the models for
links.

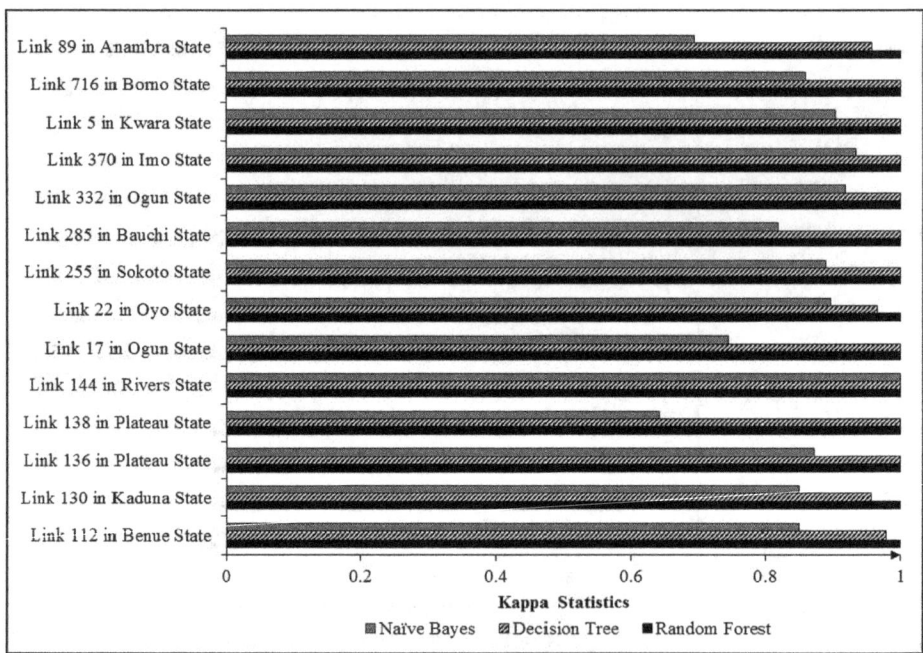

Fig. 7. Coefficient of correlation

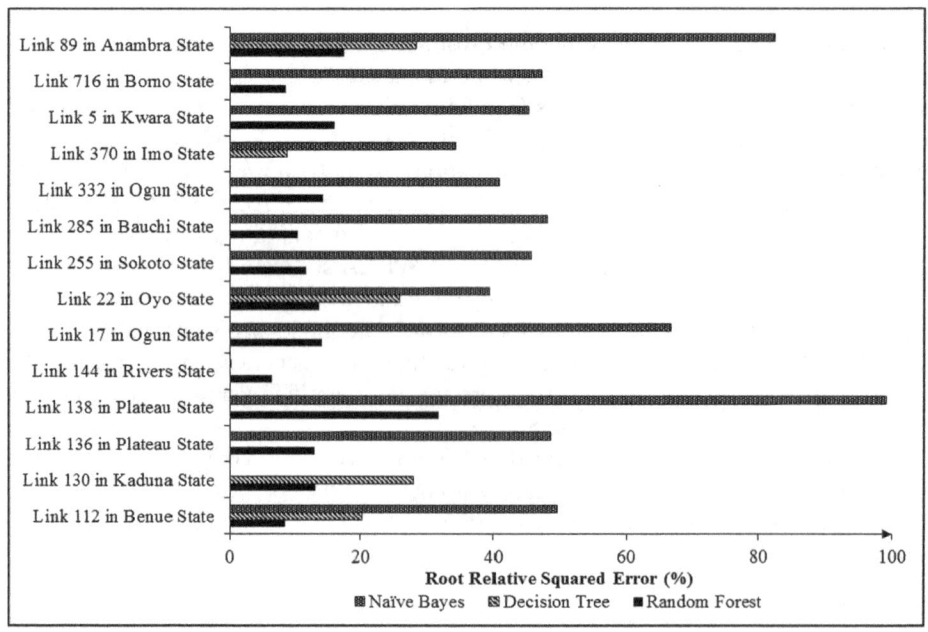

Fig. 8. Error of classification

3.2   Calibration and Validation of the FIS Model

Parameters used for the development of the FIS were adjusted to conform to realities of flexible road pavement behaviour and extreme cases of pavement conditions in Nigeria based on current conditions (Jeong *et al.* 2017; Cheu *et al.* 2004), it fitted the model inputs and outputs within limits of variables in the dataset that described the system behaviour (Chen and Flintsch 2008; Mahmood *et al.* 2013). According to Mahmood *et al.* (2013), the process requires a well spread dataset to be able to capture all possible events in the system behaviour. The process was guided by the original dataset used for model development, logical experience and engineering judgement – in other words, fitting experience to the observed logics or patterns (Chen and Flintsch 2008; Karagahin and Terzi 2014; Aggarwal and Kumar 2015; Mahmood 2015).

3.2.1   Goodness of Fit Test

Figure 9 presents a curve fitting plot of the observed and predicted results of the built FIS;

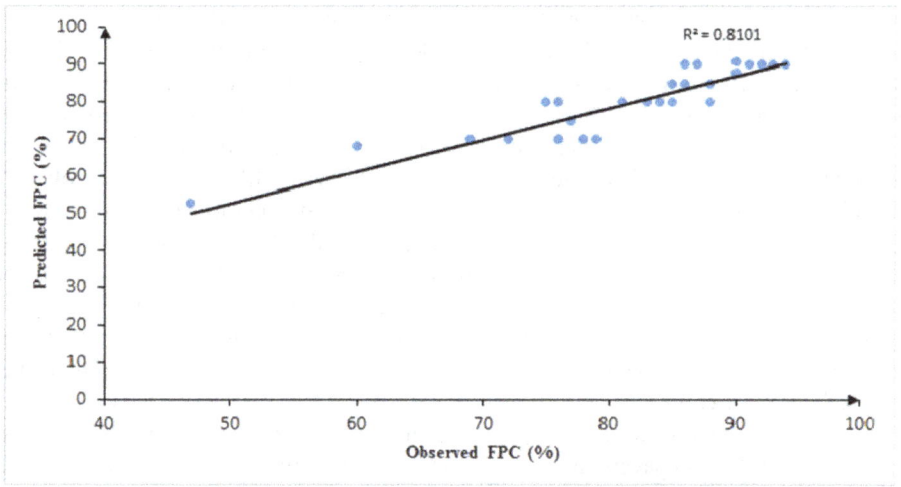

Fig. 9. Curve fitting plot

Figure 9 revealed the proportion of variance accounted for between the observed and predicted FPC values through the coefficient of determination, $R^2 = 0.8101$ with coefficient of alienation $(1 - R^2)$ at 0.1899. This defined the extent of similarity between the observed and predicted values of FPC at 81% level of accuracy. In other words, the strength of the best-fit line known as the correlation coefficient was estimated at 90% consistency (Heiman 2011).

The 'independent' dataset required for the validation of the built model was not readily available. But the relatively high degree of accuracy shown from the goodness of fit test explained the high level of certainty in using the model. Therefore, being a linguistic and experienced based model, the process adopted permutation principles to derive 5120 Fuzzy Logic Rules which satisfied possible pavement conditions in Nigeria for performance prediction (See extract in Appendix).

4 Conclusion

This study used the Random Forest, Decision Tree and Naive Bayes algorithms of data mining to examine the inferred dataset on pavement performance attributes and surface condition classification in Nigeria based on reports of pavement surface condition evaluation survey carried out in 1986. The data mining techniques were used to investigate hidden relationship between pavement performance attributes and to authenticate the accuracy of subjective measurements that were used for pavement surface condition classification. The Random Forest and Decision Tree algorithms reported relatively perfect classifications of road pavement sections into; Excellent, Good, Fair, Poor and Very poor. On the other hand, the Naïve Bayes algorithm yielded inaccurate classifications with some margin of errors which were attributed to missing and noisy entries in the dataset.

Considering outputs of the Naïve Bayes classifier, this necessitated the use of Fuzzy logic theory for performance prediction of the road pavement due to its capability to handle the imprecise dataset. It was used to develop Fuzzy Inference System (FIS) for performance prediction of flexible road pavement using attributes such as; the classified Initial Pavement Condition (IPC), Age of pavement, Resilient Modulus (M_R) of subgrade soil, Average Truck load per day, Average Annual Air Temperature and Rainfall to predict the Future Pavement Condition (FPC). The model was calibrated using the observed logical behaviour of road pavement to fit the engineering experience and judgement. A goodness-of-fit test between the observed and predicted FPC values showed high level of consistency at 90%. There were no sufficient and well-spread dataset that described pavement behavioural pattern in present Nigeria to calibrate further the Fuzzy logic rules. The process proposed 5120 mutually exclusive Fuzzy logic rules for performance prediction of road pavement based on permutation theory. Though, the required well-spread dataset for calibration of the model to cover all possible pavement conditions in Nigeria and subsequent validation were not available, a framework for performance prediction of flexible road pavement was developed, and the study presented comprehensive guidelines on how to calibrate the FIS model using well-spread dataset.

Appendix

Rule No.	Initial pavement condition (IPC)	Age (years)	Truck load (veh/day)	MR of subgrade (kg/m²)	Temp. (°C)	Rainfall (mm)	Final pavement condition (FPC)
1.	Excellent	New	Low	Low	Low	Low	Excellent
2.	Excellent	New	Low	Low	Low	Medium	Excellent
3.	Excellent	New	Low	Low	Low	High	Excellent
4.	Excellent	New	Low	Low	Low	V. High	Excellent
5.	Excellent	New	Low	Low	Medium	Low	Excellent
6.	Excellent	New	Low	Low	Medium	Medium	Excellent
7.	Excellent	New	Low	Low	Medium	High	Excellent
8.	Excellent	New	Low	Low	Medium	V. High	Excellent
9.	Excellent	New	Low	Low	High	Low	Excellent
⋮	⋮	⋮	⋮	⋮	⋮	⋮	⋮
1656.	Good	Old	Medium	V. High	Medium	V. High	Poor
1657.	Good	Old	Medium	V. High	High	Low	Poor
1658.	Good	Old	Medium	V. High	High	Medium	Poor
1659.	Good	Old	Medium	V. High	High	High	Poor
1660.	Good	Old	Medium	V. High	High	V. High	Poor
1661.	Good	Old	Medium	V. High	V.High	Low	Poor
1662.	Good	Old	Medium	V. High	V.High	Medium	Poor
1663	Good	Old	Medium	V. High	V.High	High	Poor
⋮	⋮	⋮	⋮	⋮	⋮	⋮	⋮
2370.	Fair	Recent	Medium	Low	Low	Medium	Poor
2371.	Fair	Recent	Medium	Low	Low	High	Poor
2372.	Fair	Recent	Medium	Low	Low	V. High	Poor
2373.	Fair	Recent	Medium	Low	Medium	Low	Poor
2374.	Fair	Recent	Medium	Low	Medium	Medium	Poor
2375.	Fair	Recent	Medium	Low	Medium	High	Poor
2376.	Fair	Recent	Medium	Low	Medium	V. High	Poor

(continued)

(continued)

Rule No.	Initial pavement condition (IPC)	Age (years)	Truck load (veh/day)	MR of subgrade (kg/m²)	Temp. (°C)	Rainfall (mm)	Final pavement condition (FPC)
2377.	Fair	Recent	Medium	Low	High	Low	Poor
⋮	⋮	⋮	⋮	⋮	⋮	⋮	⋮
4089.	Poor	V.Old	V.High	V. High	High	Low	V.Poor
4090.	Poor	V.Old	V.High	V. High	High	Medium	V.Poor
4091.	Poor	V.Old	V.High	V. High	High	High	V.Poor
4092.	Poor	V.Old	V.High	V. High	High	V. High	V.Poor
4093.	Poor	V.Old	V.High	V. High	V.High	Low	V.Poor
4094.	Poor	V.Old	V.High	V. High	V.High	Medium	V.Poor
4095.	Poor	V.Old	V.High	V. High	V.High	High	V.Poor
4096.	Poor	V.Old	V.High	V. High	V.High	V. High	V.Poor
⋮	⋮	⋮	⋮	⋮	⋮	⋮	⋮
5114.	V.Poor	V.Old	V.High	V. High	High	Medium	V.Poor
5115.	V.Poor	V.Old	V.High	V. High	High	High	V.Poor
5116.	V.Poor	V.Old	V.High	V. High	High	V. High	V.Poor
5117.	V.Poor	V.Old	V.High	V. High	V.High	Low	V.Poor
5118.	V.Poor	V.Old	V.High	V. High	V.High	Medium	V.Poor
5119.	V.Poor	V.Old	V.High	V. High	V.High	High	V.Poor
5120.	V.Poor	V.Old	V.High	V. High	V.High	V. High	V.Poor

References

AASHTO: AASHTO Guide for the Design of Pavement Structures. American Association of State Highway and Transportation Officials, Washington, D.C. (1993)

Abiola, O.S., Owolabi, A.O., Odunfa, S.O., Olusola, A.: Investigation into causes of premature failure of highway pavements in Nigeria and remedies. In: Proceedings of the Nigeria Institution of Civil Engineers (NICE) Conference (2010)

Abiola, O.S., Owolabi, A.O., Sadiq, O.M., Aiyedun, P.O.: Application of dynamic artificial neural network for modelling ruts depth for lagos-ibadan expressway, Nigeria. Asian Research Publishing Network (ARPN)-J. Eng. Appl. Sci. **7**(8), 986–991 (2012)

Adefemi, B.A., Ibrahim, A.A.: Flexible pavement assessment of selected highways in Ifelodun local government, Ikirun-Osun, South – Western Nigeria. Int. J. Eng. Technol. **5**(8), 475–484 (2015)

Adeke, P.T., Atoo, A.A., Joel, E.: A policy framework for efficient and sustainable road transport system to boost synergy between urban and rural settlements in developing countries: a case of Nigeria. In: 1st International Civil Engineering Conference (ICEC 2018), Department of Civil Engineering, Federal University of Technology, Minna, Nigeria (2018a)

Adeke, P.T., Atoo, A.A., Orga, S.G.: Assessment of pavement condition index: a case of flexible road pavements on the university of agriculture Makurdi Campus. Niger. J. Technol. **38**(1), 15–21 (2018b)

African Development Bank Group: Rail infrastructure in Africa – Financing Policy Options. International d'Abidjan, Abidjan, Côte d'Ivoire (2015)

Aggarwal, P., Kumar, N.: Fuzzy model for road roughness index. In: International Conference on Biological, Civil and Environmental Engineering, Indonesia, 3–4 February 2015

Ahishakiye, E., Taremwa, D., Omulo, E.O., Niyonzima, I.: Crime prediction using decision tree (J48) classificaiton algorithm. Int. J. of Comp. and Info. Techn. **6**(3), 188–195 (2017)

Alam, F., Pachauri, S.: Comparative study of J48, naïve bayes and one-r classification techniques for credit card fraud detection using WEKA. Adv. in Comp. Sci. and Tech. **10**(6), 1731–1743 (2017)

American Society for Testing and Materials (ASTM): Standard Practice for Road and Parking Lots Pavement Condition Index Survey. D6433-07, Philadelphia (2007)

Arliansyah, J., Maruyama, T., Takahashi, O.: A development of fuzzy pavement condition assessment. Proc. Jpn. Soc. Civ. Eng. **746**, 275–285 (2003)

ASTM D6433-07 2007: Standard Practice for Road and Parking Lots Pavement Condition Index Survey. American Standard for Testing and Materials, Philadelphia

Ayed, A.: Development of Empirical and Mechanistic Empirical performance Models at Project and Network Levels. PhD Thesis, Department of Civil Engineering, University of Waterloo, Canada (2016)

Bandara, N., Gunaratne, M.: Current and future pavement maintenance prioritisation based on rapid visual condition evaluation. J. Transp. Eng. ASCE **127**, 116–123 (2001)

Bianchini, A., Bandini, P.: Prediction of pavement performance through Neuro-fuzzy reasoning. Comput. Aided Civ. Infrastruct. Eng. **25**, 39–54 (2010)

Chandran, S., Isaac, K.P., Veeraragavan, A.: Prioritization of low-volume pavement sections for maintenance by using fuzzy logic. Transp. Res. Rec. J. Transp. Res. Board **1989-1**(1), 53–60 (2007). https://doi.org/10.3141/1989-06

Chassiakos, A.P.: A fuzzy-based system for maintenance planning or road pavements. In: Proceedings of the 10th WSEAS International Conference on Computers, Vouliagmeni, Athens, Greece, pp. 535–540 (2006)

Chen, C., Flintsch, G.W.: Calibrating fuzzy-logic-based rehabilitation decision models using the LTPP database. In: 7th International Conference on Managing Pavement Assets (2008)

Chen, G., Pham, T.T.: Introduction to Fuzzy Sets, Fuzzy Logic, and Fuzzy Control Systems. CRC Press, Boca Raton (2001)

Cheu, R.L., Wang, Y., Fwa, T.F.: Genetic algorithm-simulation methodology for pavement maintenance scheduling. Comput. Aided Civ. Infrastructural Eng. **19**, 446–455 (2004)

Chikezie, C.U., Olowosulu, A.T., Abejide, O.S., Kolo, B.A.: Review of application of genetic algorithms in optimization of flexible pavement maintenance and rehabilitation in Nigeria. World J. Eng. Pure Appl. Sci. **1**(3), 68–76 (2011)

Claros, G., Carmichael, R.F., Harvey, J.: Development of Pavement Evaluation Unit and Rehabilitation Procedure for Overlay Design method. Lagos: Texas Research and Development Foundation for the Nigeria Federal Ministry of Works and Housing (1986)

Cubero-Fernandez, A., Rodriguez, F.J., Villatoro, R., Olivares, J., Palomares, J.M.: Efficient pavement crack detection and classification. EURASIP J Image Video Process. **39**, 2–11 (2017)

Dabous, S.A., Al-Khayyat, G., Feroz, S.: Utility-based road maintenance prioritization method using pavement condition rating. Balt. J. Road Bridg. Eng. **15**(1), 126–146 (2019a)

Dabous, S.A., Zeiada, W., Al-Ruzouq, R., Hamad, K., Al-Khayyat, G.: Distress-based evidential reasoning method for pavement infrastructure condition assessment and rating. Int. J. Pavement Eng., 1–12 (2019b). https://doi.org/10.1080/10298436.2019.1622012

Dong, S., Zhong, J., Hao, P., Zhang, W., Chen, J., Lei, Y., Schneider, A.: Mining multiple association rules in LTPP database: an analysis of asphalt pavement thermal cracking distress. Constr. Build. Mater. **191**, 837–852 (2018)

Fox, C.: Data Science for Transport; Self-Study Guide with Computer Exercises. Springer, Switzerland (2018)

Fwa, T.F., Tan, C.Y., Chan, W.T.: Road-maintenance planning using genetic algorithms. II: analysis. J. Transp. Eng. **120**(5), 710–722 (1994)

Fwa, T.F., Shanmugam, R.: Fuzzy logic technique for pavement condition rating and maintenance-needs assessment. In: Fourth International Conference on Managing Pavements, Durban, South Africa, pp. 465–476 (1998)

Gogoi, R., Dutta, B.: Maintenance prioritisation of interlocking concrete block pavement using fuzzy logic. Int. J. Pavement Res. Technol. **13**(2020), 168–175 (2019). https://doi.org/10.1007/s42947-019-0098-9

Golroo, A., Tighe, S.L.: Fuzzy Set approach to condition assessments of novel sustainable pavements in the Canadian climate. Can. J. Civ. Eng. **36**, 754–764 (2009)

Gong, H., Sun, Y., Shu, X., Huang, B.: Use of random forest regression for predicting IRI of asphalt pavements. Const. and Building Mat. **189**, 890–897 (2018)

Gopalakrishnan, K., Agrawal, A., Ceylan, H., Kim, S., Choudhary, A.: Knowledge discovery and data mining in pavement inverse analysis. Transport **28**(1), 1–10 (2013). https://doi.org/10.3846/16484142.2013.777941

Government of The Federal Republic of Nigeria: General Specifications (Roads and Bridges), vol. II, Revised, Abuja, Nigeria (2016)

Hamed, R.I., Kakarash, Z.A.: Evaluate the asphalt pavement performance of rut depth based on intelligent method. Int. J. Eng. Comput. Sci. **5**(1), 15474–15481 (2016)

Heiman, G.W.: Basic Statistics for Behavioral Sciences, 6th edn. Cengage Learning, Belmont (2011)

Highway Research Board of the NAS-NRC Division of Engineering and Industrial Research Special Report 61G: The AASHO Road Test, Report 7 Summary Report, National Academy of Sciences – National Research Council, Washington, D.C. (1962)

Huang, Y.: Evaluating pavement response and performance with different simulative tests. PhD Thesis, Virginia Polytechnic Institute and state University, Virginia (2017)

Huang, Y.H.: Pavement Analysis and Design, 2nd edn. Pearson Prentice Hall, Inc., Upper Saddle River (2004)

Inkoom, S., Sobanjo, J., Barbu, A., Niu, X.: Prediction of the Crack condition of highway pavements using machine learning models. Struc. and Infrast. Eng. (2018). https://doi.org/10.1080/15732479.2019.1581230

Inkoom, S., Sobanjo, J., Barbu, A., Niu, X.: Pavement crack rating using machine learning frameworks: partitioning, bootstrap forest, boosted tree, Naïve Bayes, and K-Nearest neighbors. J. Transp. Eng., Part B: Pavement 145(3), 1–12 (2019)

Jang, W., Lee, J.K., Lee, J., Han, S.H.: Naïve bayesian classifier for selecting good/bad projects during the early stage of international construction bidding decisions. Math. Prob. Eng. 2015, 1–12 (2015)

Jeong, H., Kim, H., Kim, K., Kim, H.: Prediction of flexible pavement deterioration in relation to climate change using fuzzy logic. J. Infrastruct. Syst. Am. Soc. Civ. Eng. (ASCE) 23(4), 04017008-1 (2017). https://doi.org/10.1061/(asce)is.1943-555x.0000363

Karagahin, M., Terzi, S.: Performance model for asphalt concrete pavement based on the fuzzy logic approach. Transport 29(1), 18–27 (2014)

Kaur, D., Tekkedil, D.: Fuzzy expert system for ashpahl pavement performance prediction. In: IEEE Intelligent Transport Systems, Conference Proceedings, Dearborn, USA (2000)

Kirbas, U., Karasahin, M.: Performance models of hot mix asphalt pavement in urban roads. Constr. Build. Mater. 116, 281–288 (2016)

Koduru, H.K., Xiao, F., Amirkhanian, S.N., Juang, C.H.: Using fuzzy logic and expert system approaches in evaluating flexible pavement distress: case study. J. Transp. Eng. ASCE 136, 149–157 (2010)

Kudjo, P.K., Chen, J., Mensah, S., Amankwah, R., Kudjo, C.: The Effect of bellwether analysis on software vulnerability severity prediction models. Softw. Qual. J. (2020). https://doi.org/10.1007/s11219-019-09490-1

Li, Z., Chenge, C., Kwan, M., Tong, X., Tian, S.: Identifying Asphalt pavement distress using UAV LiDAR point cloud data and random forest classification. Int. J. Geo-Info. 8(39), 2–26 (2019)

Lin, J.D., Huang, W.H., Hung, C.T., Chen, C.T., Lee, J.C.: Using decesion tree for data minig of pavement maintenace and management. Appl. Mech and Mat. 330(2013), 1015–1019 (2013)

Liu, Y., Sun, M.: Fuzzy optimization BP neural network model for pavement performance assessment. In: IEEE International Conference on Grey Systems and Intelligent Services GSIS, Nanjing, China, pp. 1031–1034 (2007)

Luca, M.D., Abbondati, F., Pirozzi, M., Zilioniene, D.: Preliminary study on runway pavement friction decay using data mining. In: 6th Transport Research Area, 18–21 April 2016, vol. 14, pp. 3751–3760 (2016)

Mane, A.S., Gujarathi, S.N., Arkatkar, S.S., Sarkar, A.K., Singh, A.P.: Methodology for pavement condition and maintenance of rural roads. In: A National Conference on Fifteen Years of Transport Engineering Group, Civil Engineering Department, Indian Institution of Technology Roorkee, Roorkee – 247667 Uttarakhand (2016)

Mahmood, M.S.: Network-level maintenance decisions for flexible pavement using a soft computing-based framework. Ph.D. thesis, Nottingham Trent University, United Kingdom (2015)

Mahmood, M., Rahman, M., Nolle, L., Mathavan, S.: A Fuzzy logic approach for pavement section classification. Int. J. Pavement Res. Technol. Chin. Soc. Pavement Eng. 6(5), 620–626 (2013). https://doi.org/10.6135/ijprt.org.tw/2013.6(5).620

Mahmood, M., Rahmood, M., Mathavan, S., Nolle, L.: pavement management: data centric rules and uncertainty management in section classification by a fuzzy inference system. Bitum. Mix. Pavement 6, 533–541 (2015)

Marcelino, P. Antunes, M. de. L., Fortunato, E., Gomes, M.C.: Machine learning approach for pavement performance prediction. Int. J. Pavement Eng., 1–15 (2019). https://doi.org/10.1080/10298436.2019.1609673

Marianingsih, S., Utaminingrum, F.: Comparison of support vector machine classifier and naïve bayes classifier on road surface type classification (2018). https://doi.org/10.1109/SIET.2018.8693113

Marianingsih, S., Utaminingrum, F., Bachtiar, F.A.: Road surface types classification using combined of K-Nearest neighbor and naïve bayes based on GLCM. Int. J. Adv. Soft. Comp. Appl. **11**(2), 15–27 (2019)

MATLAB: User Guide Manual. Mathworks Inc. USA (2015)

Miradi, M.: Knowledge discovery and pavement performance: intelligent data mining, a Ph.D. thesis submitted to the Section of Road and Railway Engineering, Faculty of Civil Engineering and Geosciences, Delft University of Technology, The Netherlands (2009)

Moazami, D., Behbahani, H., Muniandy, R.: Pavement rehabilitation and maintenance prioritization of urban roads using fuzzy logic. Expert Syst. Appl. **38**, 12869–12879 (2011)

Mubaraki, M.A.: Predicting deterioration for the Saudi Arabia Urban road network. Ph.D. thesis, Department of Civil Engineering, University of Nottingham, United Kingdom (2010)

Munakata, T.: Fundamentals of New Artificial Intelligence, Neural Evolutionary, Fuzzy and More, 2nd edn. Computer and Information Science Department Cleveland State University, USA (2008)

Murana, A.A., Olowosulu, A.T., Otuoze, H.S.: Minimum threshold of monte carlo cycles for Nigerian empirical-mechanistic pavement analysis and design system. Niger. J. Technol. **31**(3), 321–328 (2012)

Murana, A.A.: Characterisation of subgrade materials from some Nigerian sources for use in the Nigeria empirical-mechanistic pavement analysis and design system, Ph.D. thesis, Department of Civil Engineering, Faculty of Engineering, Ahmadu Bello Universiy Zaria, Nigeria (2016)

Murana, A.A., Olowosulu, A.T.: Evaluation of rutting models using reliability for mechanistic-empirical design of flexible pavement. J. Eng. Appl. Sci. **7**(2), 123–127 (2012b)

Nawir, D., Prihartanto, E.: Decision-making analysis of road maintenance in north kalimantan region using technology of fuzzy logic (case study: Liang Bunyu Street Section, West Sebatik Sub-District, Nunukan District). IOP Conf. Ser. Earth Environ. Sci. **353**, 012054 (2019). https://doi.org/10.1088/1755-1315/353/1/012054

Obuandike, G.N., Audu, I., John, A.: Analytical Study of some selected classification algorithms in WEKA Using real crime dataset. Int. J. Adv. Res. Artif. Intell. **4**(12), 44–48 (2015)

Pan, Y., Zhang, X., Cervone, G., Yang, L.: Detection of asphalt pavement potholes and cracks based on the unmanned aerial vehicle multispectral imagery. IEEE J. Sel. Top. Appl. Earth Obser. Remote Sens. **1–12**, (2018). https://doi.org/10.1109/JSTARS.2018.2865528

Pantuso, A., Flintsch, G.W., Katicha, S.W., Loprencipe, G.: development of network-level pavement deterioration curves using the linear empirical Bayes approach. Int. J. Pavement Eng. (2019). https://doi.org/10.1080/10298436.2019.1646912

Pawlak, Z.: Rough sets and intelligent data analysis. Inf. Sci. **147**, 1–12 (2002)

Premkumar, L., Vavrik, W.R.: Enhancing pavement performance prediction models for the Illinois Tollway system. Int. J. Pavement Res. Technol. **9**, 14–19 (2016). https://doi.org/10.1016/j.ijprt.2015.12.002

Road Sector Development Team: Configuration and Calibration of HDM-4 to Nigeria Conditions, Government of the Federal Republic of Nigeria, Nigeria, p. 33 (2014)

Russell, S.J., Norvig, P.: Artificial Intelligence. A modern Approach, 3rd edn. Pearson Education Inc., London (2010)

Salpisoth, H.: Simple evaluation methods for road pavement management in developing countries. Ph.D. thesis - Graduate School of Engineering, Kyoto University, Japan (2014)

Saltan, M., Terzi, S., Kucuksille, E.U.: Backcalculation of pavement layer moduli and poisson's ratio using data mining. Expert Syst. Appl. **38**, 2600–2608 (2011)

Santos, J., Torres-Machi, C., Morillas, S., Cerezo, V.: A Fuzzy Logic expert system for selecting optimal and sustainable life cycle maintenance and rehabilitation strategies for road pavements. Int. J. Pavement Eng. 1–13 (2020). https://doi.org/10.1080/10298436.2020.1751161

Saravanan, N., Gayathri, V.: Performance and classification evauation of J48 algorithm and kendalls based J48 algorithm (KNJ48). Int. J. Comp. Intell. Inf. **7**(4), 188–198 (2018)

Savio, D., Nivitha, M.R., Bindhu, B.K., Krishnan, J.M.: Overloading analysis of Bituminous pavements in India using M-E pavement design guide. In: 11th Transport Planning and Implementation Methodologies for Developing Countries, Mubai, India (2016)

Setyawan, A., Nainggolan, J., Budiarto, A.: Predicting the remaining service life of road using pavement condition index. In: The 5th International Conference of Euro Asia Civil Engineering Forum (2015)

Sharma, T.C., Jain, M.: WEKA approach for comparative study of classification algorithm. Int. J. Adv. Res. Comput. Commun. Eng. **2**(4), 1925–1931 (2013)

Shekharan, A.R.: Effect of noisy data on pavement performance prediction by artificial neural networks. Trans. Res. Rec. **1643**, 7–13 (1998)

Shoukry, S.N., Martinelli, D.R., Reigle, J.A.: Universal Pavement distress evaluator based on fuzzy sets. Trans Res. Rec. **1592**, 180–186 (1997)

Smadi, O.G.: Knowledge based expert system pavement management optimisation. Paper based on Ph.D. Dissertation, Iowa State University, Ames, Iowa, United States (2000)

Surendrakuma, K., Prashant, N., Mayuresh, P.: Application of Markovian probabilistic process to develop a decision support system for Pavement maintenance management. Int. J. Sci. Technol. Res. **2**(8), 295–303 (2013)

Taylor, M.A.P., Philip, M.L.: Investigating the impact of maintenance regimes on the design life of road pavements in a changing climate and the implications for transport policy. Transp. Policy **41**, 117–135 (2015)

Thube, D.T.: Artificial Neural Network (ANN) based pavement deterioration models for low volume roads in India. Int. J. Pavement Res. Technol. **5**(2), 115–120 (2011)

Transport Research Board of the National Academies: Transport Research Circular E-C118 - Pavement Lessons Learned from the AASHO Road Test and Performance of the Interstate Highway System, Washington, DC 20001 (2007)

Tribhuvan, A.P., Tribhuvan, P.P., Gade, J.G.: Applying naïve bayesian classifier for predicting performance of a student using WEKA. Adv. Comp. Res. **7**(1), 239–242 (2015)

Uglova, E.V., Tiraturyan, A.N.: Interlayer bond evaluation in the flexible pavement structures using a non-destructive testing method. In: International Conference on Industrial Engineering (2016)

Wang, K.C.P., Li, L.: Pavement smoothness prediction based on fuzzy and gray theories. Comput. Aided Civ. Infrastruct. Eng. **26**(1), 69–76 (2011)

WEKA: Waikato Environment for Knowledge Analysis: User's Manual, The University of Waikato, New Zealand (2018)

Witten, I.H., Frank, E.: Data Mining – Practical Machine learning Tools and Techniques, 2nd edn. Morgan Kaufmann, Elsevier, London (2005)

Witten, I.H., Frank, E., Hall, M.A., Pal, C.J.: The WEKA Workbench – Data Mining Practical Machine Learning Tools and Techniques, 4th edn. Morgan Kaufmann, Elsevier, London (2016). Online Appendix, Retrieved on 6th February, 2020. From: www

Witten, I.H., Frank, E., Hall, M.A., Pal, C.J.: The WEKA Workbench – Data Mining Practical Machine Learning Tools and Techniques, 4th edn. Morgan Kaufmann, Elsevier, London (2016)

Woo, S., Yeo, H.: Optimization of pavement inspection schedule with traffic demand prediction. In: 11th International Conference of the International Institute for Infrastructure Resilience and Reconstruction (13R2) (2016). Procedia – Social and Behavioural Sciences 218, 95–103

Xiao, D.X., Wu, Z.: Using systematic indices to relate traffic load spectra to pavement performance. Int. J. Pavement Res. Technol. **9**, 302–312 (2016)

Yin, H.: Integrating instrumentation data in probabilistic performance prediction of flexible pavement, Ph.D. thesis in the Department of Civil and Environmental Engineering, Graduate School, The Pennsylvania State University (2007)

Yu, J.: Pavement service life estimation and condition prediction. Ph.D. thesis, Department of Civil Engineering, University of Toledo (2005)

Ziliute, L., Motiejunas, A., Kleiziene, R., Bribulis, G., Kravcovas, I.: Temperature and Moisture Variation in Pavement Structures of the Test Road. 6th Transport Research Arena (2016)

Design and Detailing of Bridge Approach Slabs: Cast-in-Place and Precast Concrete Options

Mostafa Abo El-Khier and George Morcous[✉]

University of Nebraska-Lincoln, 1110 S. 67th Street, Omaha, NE 68182-0816, USA
maboel-khier2@huskers.unl.edu, gmorcous2@unl.edu

Abstract. Approach slab is a structural concrete slab that spans from the back wall of the abutment (i.e. end of the bridge floor) to the beginning of the paving section. The purpose of the approach slab is to carry the dead and live loads over the backfill behind the abutments to avoid differential settlement that causes bumps at the bridge ends. Cast-in-place (CIP) concrete approach slab is the current practice in most of the states in US with various spans, reinforcement, thicknesses, and concrete covers. However, it has been reported that most approach slabs experience cracking and settlement, which result in premature deterioration and shorter service life. The replacement of deteriorated approach slabs causes costly and long traffic closure and detouring. Precast concrete (PC) approach slabs is a promising solution that could provide longer service life and accelerated construction/replacement. This paper presents a literature on current approach slab practices and innovative precast concrete solutions. Also, an analytical investigation is conducted using finite elements to evaluate the performance of the current approach slab practices in the state of Nebraska. Several parameters are considered in this investigation, such as volume changes due to shrinkage and temperature changes as well as skew angle and bridge width. Analysis results indicate that volume changes cause high tensile stresses along abutment line, which results in longitudinal cracks. Also, high skew angles result in stress concentrations at the slab corners and the increase in slab width increases the stresses in transverse direction.

1 Introduction

Approach slabs are usually supported by the back wall of the abutment at one end and a grade beam or sleeper slab at the other end. Soil backfill supports the approach slab in between the two ends across the bridge width. Figure 1 shows the plan view of a typical approach slab system. Despite the simplicity of approach slab structural system and its design as a one-way cast-in-place (CIP) reinforced concrete slab, it has been reported that most approach slabs experience cracking at early ages, which results in premature deterioration and shorter service life. The purpose of the paper is to present the different practices of approach slab design and detailing according to several US Departments of Transportation (DOTs). Special attention will be given to the current practice of Nebraska Department of Transportation (NDOT) and its recent implementation of precast concrete (PC) approach slabs as alternative to CIP concrete approach

H. Shehata and S. El-Badawy (Eds.): *Sustainable Issues in Infrastructure Engineering*, SUCI, pp. 193–206, 2021.
https://doi.org/10.1007/978-3-030-62586-3_12

slabs to minimize deterioration and construction duration. A finite element analysis was also conducted to evaluate the behavior of approach slabs under dead load, live load, and volume changes. Other parameters were considered in this investigation including skew angle and bridge width.

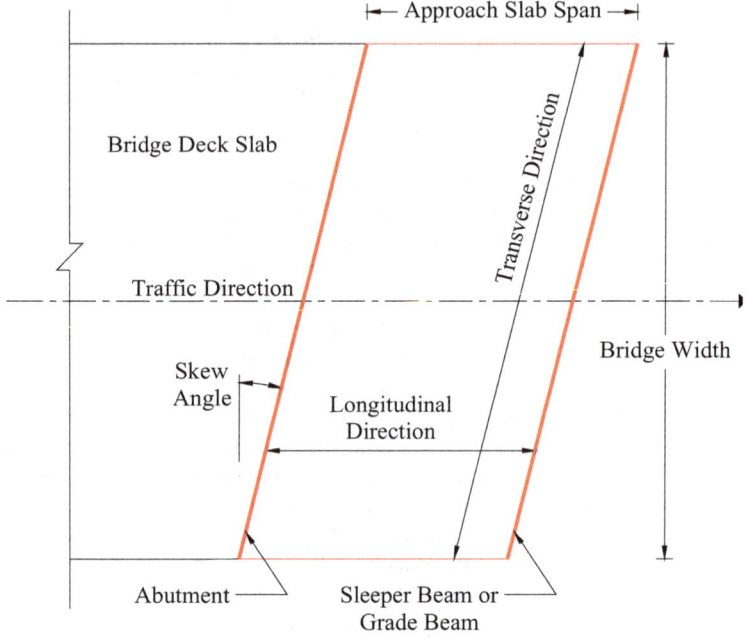

Fig. 1. Typical approach slab system

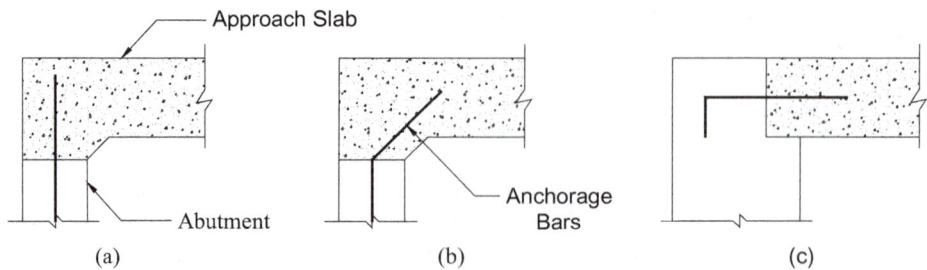

Fig. 2. Different anchorage bar layouts at joint between approach slab and abutment; (a) vertical, (b) bent, and (c) horizontal.

2   Current Practice of Bridge Approach Slabs in US

The current practice of CIP concrete approach slabs in US vary among State DOTs with respect to the following parameters: slab length, slab thickness, concrete cover, and top and bottom longitudinal and transverse reinforcement. Thiagarajan *et al.* (2010) performed a comprehensive review of approach slab practices in US states DOTs to develop approach slab design and detailing recommendations and perform cost analysis. Below is a summary of ranges of the different design parameters:

1. Span length ranges from 10 ft. to 33 ft.
2. Slab thickness ranges from 8 in. to 17 in.
3. Concrete cover ranges from 1 in. to 4 in.
4. Bottom longitudinal reinforcement ranges from #5 @ 8 in. to #10 @ 6.5 in.
5. Top longitudinal reinforcement ranges from #4 @ 18 in. to #7 @ 12 in.
6. Bottom transverse reinforcement range from #4 @ 24 in. to #6 @ 6 in.
7. Top transverse reinforcement ranges from #4 @ 18 in. to #6 @ 12 in.

Table 1 summarizes the current approach slab detailing for five different U.S. states that represent different geographic and climatic regions: California Department of Transportation (Caltrans); Washington State Department of Transportation (WsDOT); Missouri Department of Transportation (MoDOT); Iowa Department of Transportation (Iowa DOT); and Colorado Department of Transportation (CDOT).

Table 1. Different approach slab designs

State DOT		Caltrans (Type N)	WSDOT	MoDOT	Iowa DOT	CDOT
Span (ft.)		30	25 (at short edge)	20	20 (centerline)	20
Slab thickness (in.)		14	13	12	12	12
Main longitudinal reinforcement	Top	#5 @18″	#6@5″	#5@12″	#6@12″	#4 @18"
	Bottom	#10 @6″	#8@5″	#6 @5″	#8 @12″	#6 @6″
Concrete cover (in.)	Top	2	2.5	2	2.5	3
	Bottom	2	2	2	2.5	3
Transverse reinforcement	Top	#5@18″	#5 @18″	#5 @12″	#5 @12″	#5 @12″
	Bottom	#5 @6″	#5@9″			
Abutment joint type (Fig. 2)		Vertical #5@9″	45° bent #5@12″	Horizontal #5 @12″	Vertical Stainless-Steel Dowel	Horizontal #5 @12″
Other edge joint type		Horizontal #6 dowel @12″ to paving	Horizontal 1.5 in. diameter dowel bar @ 18 in. to paving	Resting on sleeper slab		

Chee (2018) conducted a survey covering 23 US DOTs, including NDOT, located in middle to east coast of US to collect data about the performance of existing approach slabs. The survey covered the primary issues with approach slab, cracking direction and location, and methods to minimize approach slab cracking. Most of the surveyed DOTs stated that settlement and concrete cracking are the top two problems with approach slabs. Transverse and longitudinal cracks are the common crack patterns in most states while few states reported diagonal cracking in addition. The methods recommended to minimize approach slab cracking include increasing thickness and/or reinforcement; limiting slab dimensions by adding joints; treating approach slabs similar to decks with respect to curing; and using sleeper slab with piles as shown in Fig. 3.

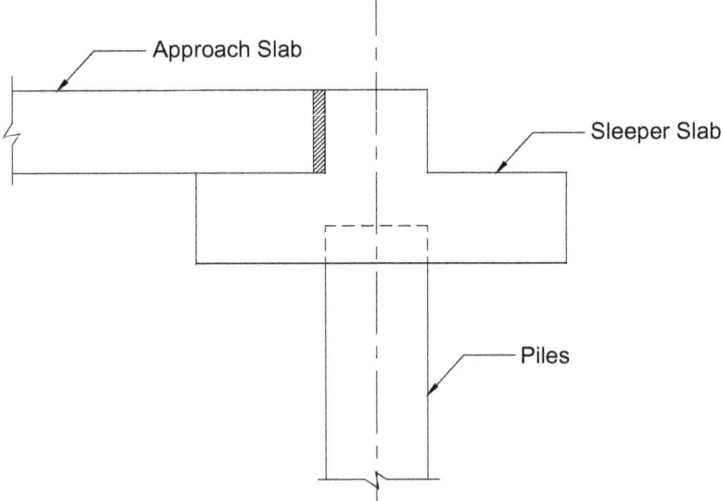

Fig. 3. Sleeper slab with piles.

3   Nebraska Department of Transportation (NDOT) Practice

A 14 in. approach slab is specified by NDOT to be simply supported by the abutment and the grade beam as shown in Fig. 4 (BOPP 2016). The grade beam is a reinforced concrete beam parallel to the abutment, supported by piles to minimize settlement, and extended to cover sidewalk. The minimum span length of approach slab is 20 ft. measured at the centerline of roadway from the end of bridge floor to centerline of grade beam. The main longitudinal reinforcement is #8 @ 5 in. and #5 @ 12 in. for bottom and top reinforcement, respectively. The transverse reinforcement is #5 @ 12 in. and #5 @ 8 in. for top and bottom reinforcement, respectively. The main longitudinal reinforcement cover is 2.5 in. and 3 in. for top and bottom reinforcement, respectively. The approach slab is anchored to the abutment using #6 bar bent at 45 deg. inside the approach slab with adequate development length and spaced at 12 in.

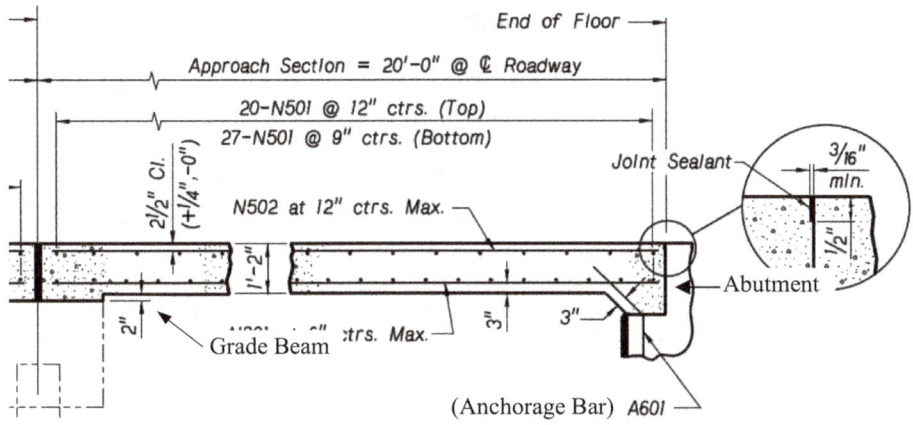

Fig. 4. Current NDOT bridge approach slab (BOPP 2016)

Figure 5 shows the approach slab cracking patterns in two bridges in Nebraska. The figure shows several longitudinal cracks extending from the abutment and grade beam ends towards the middle of approach slab.

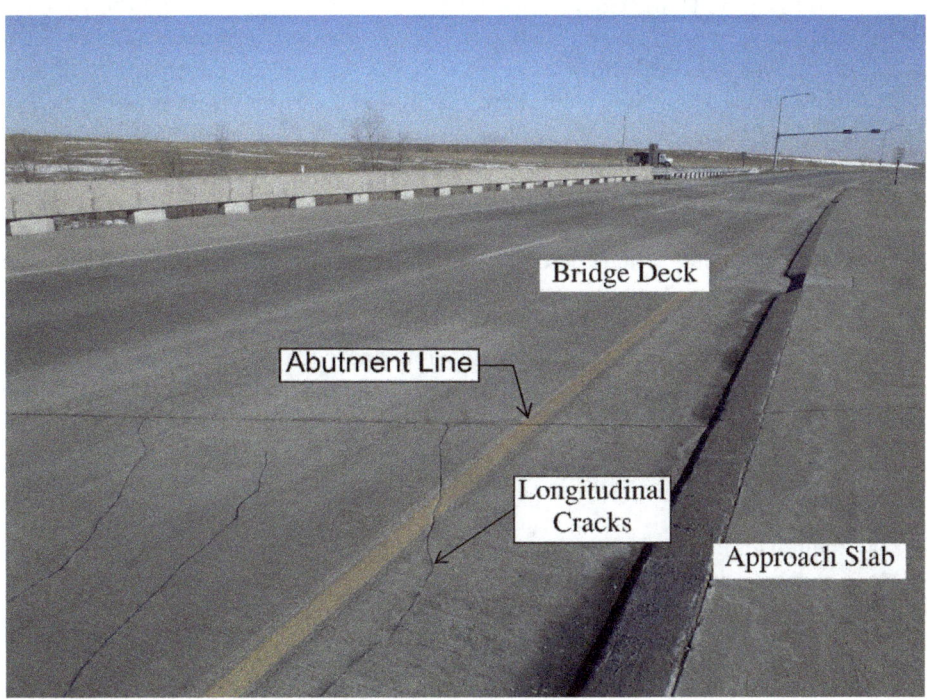

Fig. 5. Example of approach slab cracking in Nebraska

4 Precast Bridge Approach Slab

CIP concrete approach slabs face several challenges, such as unexpected weather conditions and low-quality control during placing and curing. These challenges could severely affect the performance of CIP concrete approach slabs and eventually lead to cracking. Precast concrete approach slabs minimizes these challenges as they are fabricated in a controlled environment under high-quality control that ensures reaching the desired properties. Merritt et al. (2007) reported the replacement of the approach slab of a bridge on Highway 60 near Sheldon, Iowa by eight precast concrete panels with dimensions of 20 ft. × 14 ft. × 12 in. The precast slabs were post-tensioned in both direction using 0.6 in. grade 270 7-wire stand @ 24 in. and a flowable grout was used to fill the ducts. The slab placement started from the bridge abutment after fitting #8 stainless steel anchorage bars in sleeves formed in precast panels. Each precast panel had #8@12 in. and #6@24 in. as bottom and top longitudinal reinforcement, respectively, and #5@12 in. for top and bottom transverse reinforcement. A key-shape transverse joints were used to connect the panels using epoxy after aligning the longitudinal post-tension ducks. However, the longitudinal joint was filled with grout. The under-slab was filled by pumped grout. The construction faced some challenges such as aligning panels with skewed bridge floor and post-tension ducts, and panel end damage during post tensioning. Four 12 in. thick precast concrete slabs were used in a replacement bridge over Big Brown Creek on River Road (S-86) in Union County, South Carolina. The bridge was 37.25 ft. wide and had a skew angle of 38°. These panels were tested and long-term monitored by Ziehl et al. (2015). First, the back-fill was replaced by #789 stone and cover by a 6 in. thick roller compacted macadam as a sub-base material and polyethylene moisture barrier. Then, a CIP ledger was cast with vertical dowel bars. The approach slab panels were placed, starting from exterior panel, after filling the anchorage dowels in formed sleeves. A grout was used to fill the dowel sleeves after installing the panels. The longitudinal joints between panels had longitudinal 2#6 bars tied to top and bottom of overlapped #5-U shaped bars and filled with concrete. Separation cracks were noticed at the abutment joint in the 2.5 in. thick asphalt layer placed over approach slab.

In 2012, Precast/Prestressed Concrete Institute (PCI) published guidelines presenting suggested design and details for precast concrete approach slabs. Two typical precast designs were presented simulating two cases: surface approach slab and sub-surface approach slab, as shown in Fig. 6. Also, the guidelines contain different joint configurations for longitudinal and transverse directions. Below are the requirements for using the proposed designs and details:

- Maximum width of 12 ft. for each panel including any projecting reinforcement
- Maximum weight of 100 kips
- Minimum concrete compressive strength 5,000 psi
- Using shrinkage compensating admixture for site cast concrete
- Grout is used for small voids (flowable, same strength of concrete)

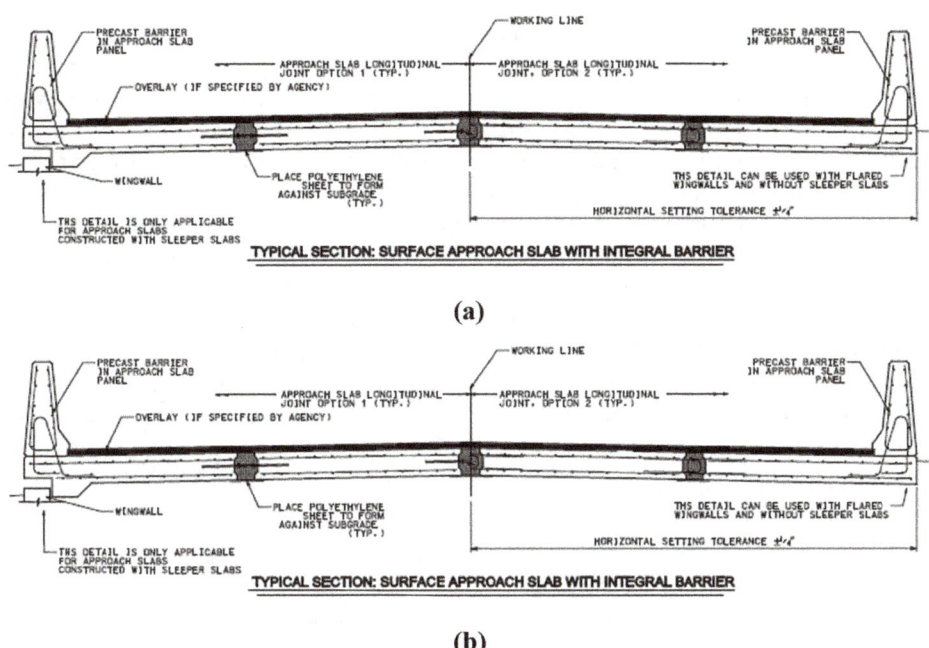

Fig. 6. PCI precast approach slab; (a) surface approach slab and (b) sub-surface approach slab (PCI 2012)

Fig. 7. Precast concrete approach slabs in Nebraska; (a) Belden-Laurel Bridge Project and (b) I-680/West Center Bridge Project

NDOT had successfully implemented the precast approach slab concept in two projects. The first project is replacing the Belden-Laurel bridge on U.S. 20 over Middle Logan Creek in Cedar County, NE in 2018. The project was the first bridge constructed entirely using prefabricated components, including approach slabs, for accelerated bridge construction in Nebraska. The bridge width was 42 ft. 8 in. Four approach slab panels were used to construct each approach slab of the bridge. Longitudinal joints filled with High Early Strength Concrete (HESC) connected the precast panels and then Ultra-High Performance Concrete (UHPC) were used to fill the transverse joint between panels and bridge deck as shown in Fig. 7. Finally, Flowable fill was pumped underneath the panels to fill the gaps between the panels and backfill. The second project is the replacement of I-680/West Center Road Bridge. The replacement was conducted in two stages, each stage replaced half of the approach slabs using three precast concrete panels and precast concrete rail. Panels were prefabricated by the contractor at his yard and transported and placed overnight road closure. Reinforced longitudinal joints were filled with HESC and vertical dowels bars were used to connect the panels to the abutment using 3 in. diameter dowel holes, while horizontal tie bars were used to connect the panels to the paving section. Figure 8 shows construction joints between precast concrete approach slabs and between approach slab and abutment used in Nebraska.

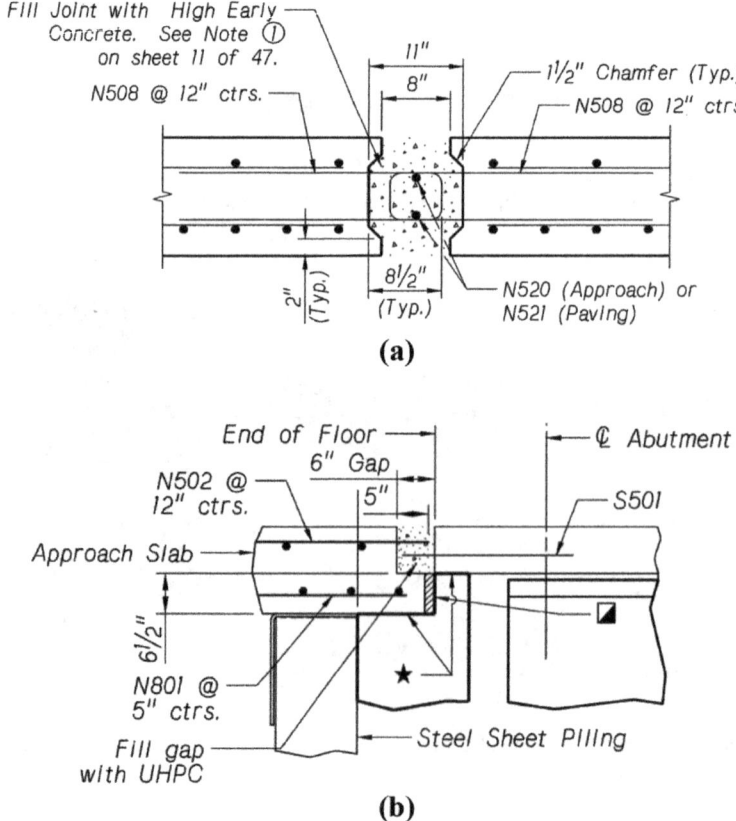

Fig. 8. Construction joints between precast concrete approach slabs in Nebraska; (a) longitudinal joint and (b) transverse joint

5   Analytical Investigation

A parametric study was conducted using Ansys V19 R1 to create a finite element model (FEM) simulating the current practice of approach slab in Nebraska. The properties of an existing bridge in Nebraska were used to create the FEM. The bridge had a skew angle of 14° and the approach slab was 14 in. thick, 43 ft. wide, and 20 ft. span. FEM was used to investigate several parameters including skew angle, volume changes, and approach slab width. The parameters considered in this investigation are shown in Fig. 9. According to BOPP Manual 2016, the required compressive strength for approach slab is 4000 psi., therefore, the cracking stress (modulus of rupture) of normal weight concrete was estimated to be 474 psi according to AASHTO LRFD (2017).

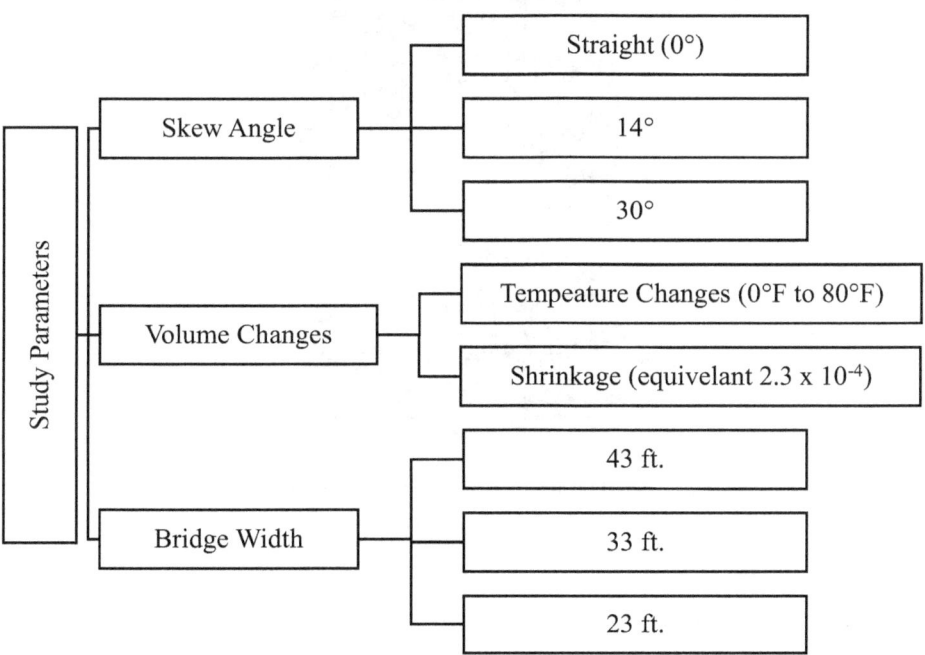

Fig. 9. Parameters considered in the study

The approach slab was simulated using *Solid65* element with an initial dimension as shown in Fig. 10. *Solid65* element allows to simulate slab thickness and rebars, define the cracking and crushing limits, and add the nonlinear material properties. The elements were meshed to a maximum size of 1 ft. to obtain accurate results and parallel to the abutment and slab edge. The joints between approach slab and abutment were simulated as hinge supports every 1 ft. The connection between the approach slab and grade beam was simulated with roller support as there is no anchorage bars and there is no restriction on approach slab horizontal movement. The own weight of the slab was considered in all the cases. The wheel loads were applied as a pressure tire covering 12 in. × 24 in. to fit with the meshing size, which is slightly larger than the 10 in. × 20 in. specified in

AASHTO LRFD (2017). The FEM was solved in material linear behavior to obtain the cracking stresses.

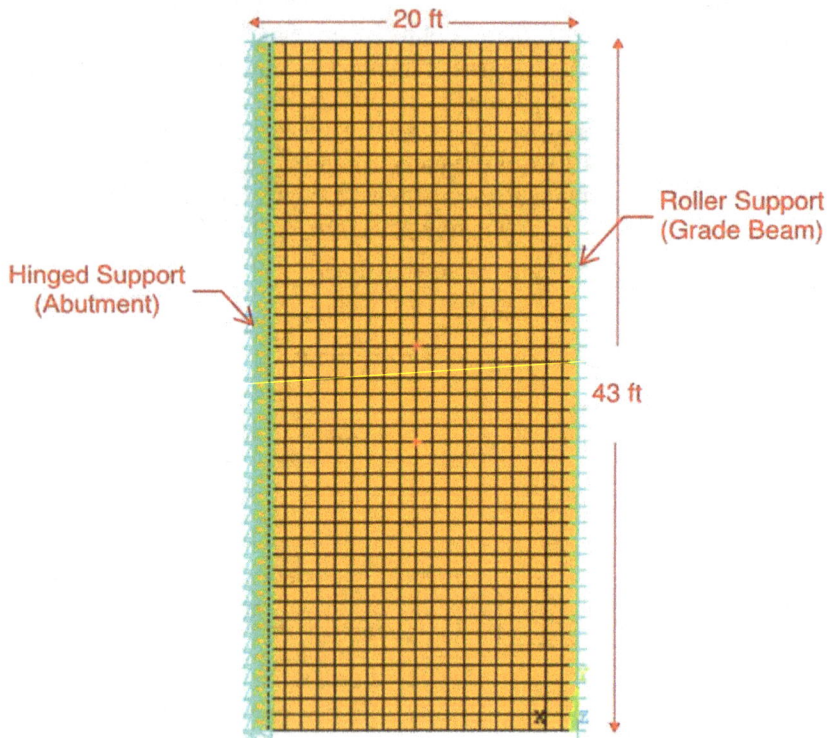

Fig. 10. Finite element model using Ansys V19.

5.1   Skew Angle Effect

Three different bridge skew angles were investigated with placing tandem axle load with impact at 2 ft from the approach slab edge. Figure 11 shows the principle tensile stresses at the top surface of slab due to bridge skew angle. The principle tensile stresses increase at the slab corner with the increase of the skew angle. However, these stresses did not exceed the concrete cracking stress.

5.2   Volume Changes Effect

The effect of volume changes on concrete approach slab was investigated for both concrete shrinkage and uniform temperature change conditions. The concrete shrinkage was calculated according to AASHTO LRFD (2017) Section 5.4.2.3.3 and was determined to be 2.30×10^{-4} at 28 days for the current NDOT curing practices. The temperature changes was determined by AASHTO LRFD (2017) Section 3.12.2.1 procedure A,

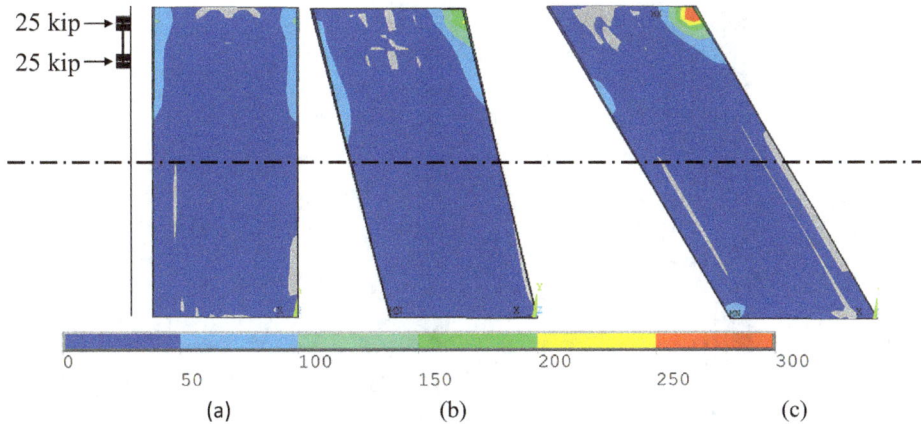

Fig. 11. Effect of skew angle on principle tensile stress at slab top surface (psi); (a) straight (0°), (b) 14°, and (c) 30°

which requires a temperature change from 0 °F to 80 °F. A coefficient of thermal expansion of 6.0×10^{-6} in./in./°F was used for concrete approach slab. Based on the analysis, the uniform temperature change was more critical than concrete shrinkage and, therefore, its effect was investigated for different bridge skew angles. Figure 12 shows the principle tensile stresses at the top surface of slab due to uniform temperature change, which are primarily concentrated along the abutment support line. The obtained tensile stresses are higher than the concrete modulus of rupture and, therefore, results in the observed longitudinal cracking. Figure 13 shows the directions of tension stresses on the top surface of approach slab that explains the cracking phenomena along abutment line.

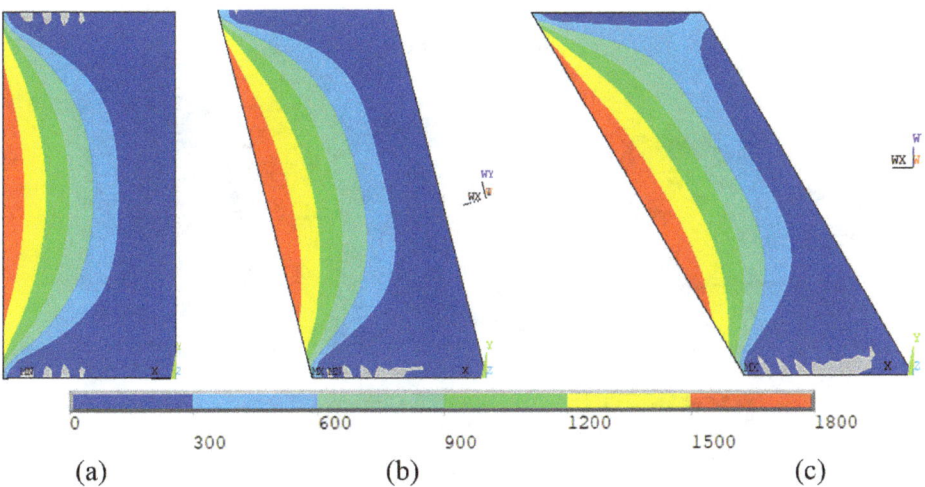

Fig. 12. Effect of concrete shrinkage on principle tensile stress at slab top surface (psi); (a) straight (0°), (b) 14°, and (c) 30°

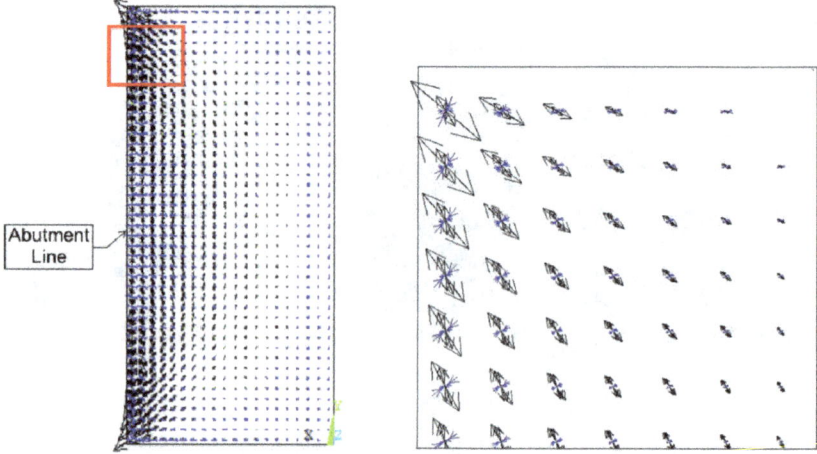

Fig. 13. Directions of tensile stresses at the top surface of approach slab along abutment line

5.3 Bridge Width Effect

The effect of different bridge width on approach slab performance was studied on 14°
skewed bridge case. A 23 ft., 33 ft., and 43 ft. approach slab widths were applied to
FEM to obtain the maximum stresses in both directions, longitudinal and transverse, at
the bottom surface of approach slab. It was found that applying the tandem axle load at
the middle of approach slab gave more representative clarification for the bridge width
effect. Figure 14 and 15 show the effect of bridge width on longitudinal and transverse
stresses at slab bottom surface. Increasing the approach slab width redistributes the

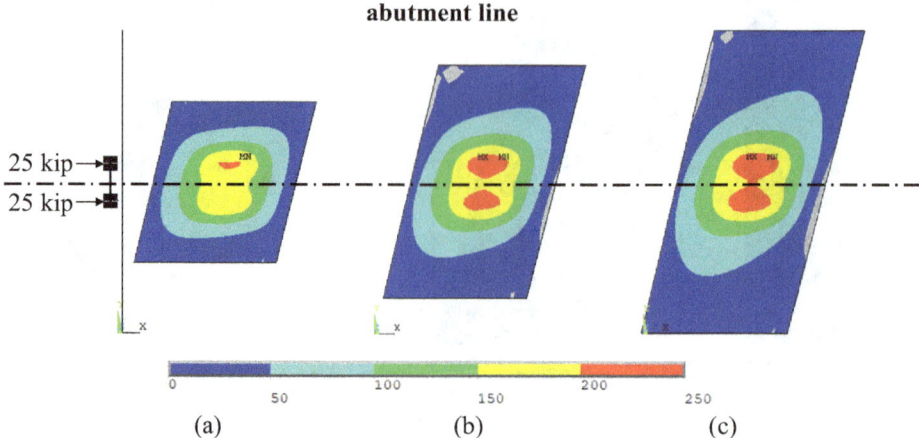

Fig. 14. Effect of bridge width on tensile transverse stresses at slab bottom surface (psi); (a) 23
ft., (b) 33 ft., and (c) 43 ft.

stresses due to wheel load between longitudinal and transverse directions as shown in Fig. 15. This figure shows that the longitudinal stress decrease by 11% and the transverse stress increases by 14.6% when the slab width increases from 23 ft. to 43 ft. So, it can be concluded that the transverse stresses increase with the increase of slab width for the same bridge skew angle which need to be considered in the design.

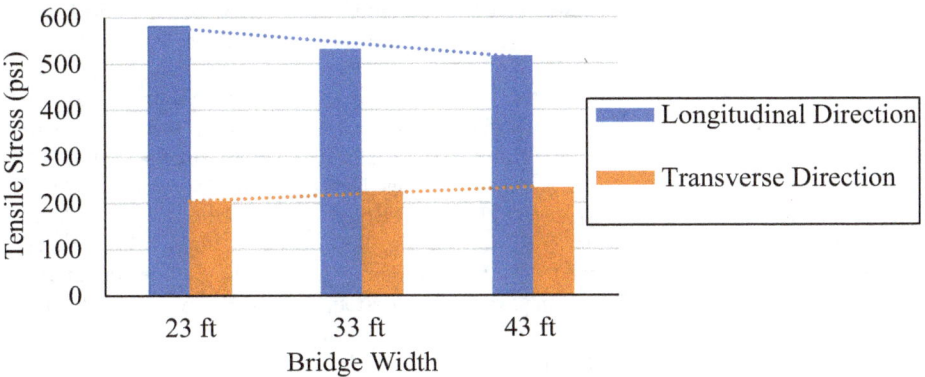

Fig. 15. Effect of bridge width on the tensile stresses' distribution at slab bottom surface for 14° skewed bridge.

6 Conclusions

This paper presents a literature review on the current practices of approach slab in Nebraska and the other DOTs. The causes of approach slab deterioration and its possible solutions were discussed. Also, a parametric study was conduction by finite element modelling. The following conclusions were drawn from this study:

1. The current practices for design and detailing of CIP approach slabs has different parameters with respect to slab length, slab thickness, concrete cover, construction joints and top and bottom longitudinal and transverse reinforcement.
2. Concrete cracking and differential settlement are the two common issues with approach slabs. Using grade beam resting on piles, which is the practice of NDOT, eliminates the settlement problem, however, longitudinal cracking is still a concern.
3. High skew angles result in concentrated tensile stresses at slab top surface at corners under live load.
4. Volume changes due to shrinkage and temperature generates high tensile stresses along abutment line that exceeds the concrete cracking stress. The direction of these stresses explains the cracking phenomena of approach slab along abutment line.
5. The transverse stresses increase with the increase of approach slab width for the same bridge skew angle, which needs to be considered in the design while the longitudinal stresses decrease.

Acknowledgements. The presented work was funded by the Nebraska Department of Transportation (NDOT) [SPR-1(19) (M085)].

References

American Association of State Highway and Transportation Officials (AASHTO): LRFD Bridge Design Specifications. AASHTO LRFD, 8th edn. AASHTO, Washington, D.C. (2017)

ANSYS Manual, A. F. U.: Release 16.0. ANSYS Inc. (2016)

California Department of Transportation (Caltrans): Standard Plans Book. California, USA (2018)

Chee, M.M.W.: Assessment of structural concrete approach slab cracking at integral abutment bridges. Masters' Thesis. University of Illinois at Urbana-Champaign, Urbana, Illinois (2018)

Colorado Department of Transportation (CDOT): Bridge Design Manual. Colorado, USA (2020)

Iowa Department of Transportation (Iowa DOT): LRFD Bridge Design Manual, Iowa, USA (2020)

Merritt, D.K., Miron, A.J., Rogers, R.B., Rasmussen, R.O.: Construction of the Iowa Highway 60 Precast Prestressed Concrete Pavement Bridge Approach Slab Demonstration. Report No. FHWA-IF-07-034, US Department of Transportation, Federal Highway Administration, Washington, D.C. (2007)

Missouri Department of Transportation (MoDOT): Bridge Standard Drawings, Approach Slabs - APP. Missouri, USA

Nebraska Department of Transportation (NDOT): Bridge Office Policies and Procedures (BOPP) Manual. Nebraska, USA (2016)

Precast/Prestressed Concrete Institute (PCI): Guideline for Precast Approach Slab (2012). https://www.pci.org/PCI_Docs/PCI_Northeast/Technical_Resources/Bridge/2013_05_29%20App%20Slab.pdf

Thiagarajan, G., Gopalaratnam, V., Halmen, C., Ajgaonkar, S., Ma, S., Gudimetla, B., Chamarthi, R.: Bridge Approach Slabs for Missouri DOT: Looking at Alternative and Cost Efficient Approaches. Report No. OR11.009, Missouri Department of Transportation, USA (2010)

Washington State Department of Transportation (WSDOT): Design - Standard Plans Manual. Washington State, USA (2019)

Ziehl, P., ElBatanouny, M., Jones, M.K.: In-situ Monitoring of Precast Concrete Approach Slab Systems. South Carolina Department of Transportation, USA (2015). https://www.scdot.scltap.org/wp-content/uploads/2015/03/SCDOT-Precast-Approch-Slabs-Final-Report.pdf

Exploring the Durability Specification of Coarse Aggregate Used in Airport Asphalt Mixtures

Greg White[(✉)] and Kayl Fergusson

University of the Sunshine Coast, Sippy Downs, QLD, Australia
gwhite2@usc.edu.au

Abstract. Coarse and fine aggregate constitutes approximately 93% of dense graded airport asphalt and the aggregate properties can affect asphalt surface performance. Despite a general trend towards performance-related specification of asphalt mixtures, prescriptive aggregate properties are generally still retained. This primarily reflects the absence of reliable performance-based laboratory test methods for determining the effect of aggregates on asphalt weathering and erosion. Historical airport asphalt specifications included a broad range of aggregate durability properties and the aggregate supply industry has questioned whether coarse aggregate durability testing can be simplified to combinations of just two properties. To determine whether a reduction in aggregate durability testing is appropriate for Australian airport asphalt, eight sources of aggregate were tested for wet strength, wet-dry strength variation, Los Angeles abrasion, sodium sulphate soundness and water absorption. The different tests were associated with different levels of variability and the correlation between the various tests results was generally low, except for Los Angeles abrasion and wet strength. The industry recommended combinations of aggregate durability testing were found to be inconsistent and ineffective. Consequently, the current range of aggregate durability tests must be retained. The only exception was the potential to omit Los Angeles abrasion when the wet strength is high. Furthermore, there was no significant difference between the results associated with the various coarse aggregate fraction sizes, indicating it may be appropriate to allow only one sized fraction per quarry source to be tested. Further work is required to correlate the various aggregate durability tests to asphalt field performance.

1   Introduction

Flexible airport pavement surfaces are predominantly comprised of dense graded, Marshall designed asphalt. The aggregates are usually fully crushed, newly quarried hard rock and the bituminous binder is usually a premium or modified product. The binder content is high compared to road and highway asphalt, typically 5.4–5.8% by mass of the asphalt mixture. Some jurisdictions and airports have developed alternate airport surface types. However, the USA, Australia, New Zealand, the Middle East and South Africa continue to favour dense graded mixtures designed using the Marshall method (White 1985).

Regardless of the asphalt mixture type, it is clear that asphalt surface performance is critical to the lifecycle cost and efficiency of airport pavement systems (AAA 2017).

© The Author(s), under exclusive license to Springer Nature Switzerland AG 2021
H. Shehata and S. El-Badawy (Eds.): *Sustainable Issues in Infrastructure Engineering*, SUCI, pp. 207–224, 2021.
https://doi.org/10.1007/978-3-030-62586-3_13

Generally, the performance requirements for airport asphalt are similar to those associated with road and other pavement surfaces. However, the prioritisation of the necessarily balanced performance properties is skewed to reflect the slow and unstable movement of aircraft on the ground, the importance of skid resistance during take-off and landing operations, as well as the potential for loose stones to damage fragile aircraft engines (Table 1).

Table 1. Airport asphalt performance requirements (White 2018)

Physical requirement	Protects against	Importance
Deformation resistance	Groove closure Rutting Shearing/shoving	High
Fracture resistance	Top down cracking Fatigue cracking	Moderate
Surface friction and texture	Skid resistance Compliance requirement	High
Durability	Pavement generated loose stones Resistance to moisture damage	Moderate

Of the airport asphalt performance properties identified in Table 2, deformation resistance, fracture resistance and surface texture can be tested in the laboratory. Established test methods are directly related to the asphalt mixture performance in the field and these tests form the basis of performance-related specifications (White 2018). For example, wheel tracking at 60 °C is well established as being related to the risk of asphalt shearing, shoving, rutting and groove closure in the field (Jamieson and White 2019). Indirect tests are also well established for moisture damage resistance, such as the loss of indirect tensile modulus upon vacuum soaking asphalt samples with high air voids, known as the Lottman test (Nosler and Beckedahl 2000). In contrast, testing for durability associated with fretting, ravelling and the generation of loose stones (USACE 2009) is not well established.

Fretting, ravelling and other distresses that contribute to pavement generated loose stones are generally a function of age related weathering. The factors that affect the weathering of asphalt surfaces include (Abouelsaad and White 2020):

- Binder properties. The propensity for a particular bituminous binder to harden with age due to oxidation directly effects on embrittlement of the bituminous mastic and the rate of asphalt surface weathering.
- Aggregate properties. The propensity for particular aggregates to absorb the bituminous binder, or to degrade and breakdown, directly effects the loss of the mastic and the rate of weathering.
- Mixture composition. The volumetric composition of the mixture affects the amount and composition of the mastic and the bituminous binder film thickness, consequently affecting the amount of mastic erosion that occurs before ravelling commences.

Table 2. Typical Australian airport asphalt characteristics (Emery 2005)

Asphalt property	Typical value
Bitumen content (% by mix mass)	5.8
Target Marshall air voids (%)	4
Target voids filled with bitumen (%)	75
Filler Content (% by aggregate)	1.5 of hydrated lime
Minimum Marshall Stability (kN)	12
Maximum Marshall Flow (mm)	3
Percentage passing AS sieve (mm)	Target by volume (%)
13.2	100
9.5	82
6.7	70
4.75	60
2.36	44
1.18	33
0.600	25
0.300	16
0.150	10
0.075	5

Construction factors also play a part with sound joint construction and protection from mixture segregation required to reduce the rate of severe weathering in isolated areas. Although significant research on asphalt weathering has focussed on binder properties and the mixture composition, the aggregate properties also play an important role. In particular, the durability of the aggregate is expected to be important and because there is no established performance-related test for asphalt mixture weathering associated with aggregate durability, aggregate durability is generally still specified in a prescriptive manner.

This paper reviews the specification of aggregate durability for airport asphalt mixtures, particularly within the Australian context. A range of diverse aggregate samples were tested for the various aggregate durability properties and the results are explored to consider whether the testing indicates similar or different relative aggregate suitability. The conclusions consider whether there is an opportunity to reduce or omit some, or all, of the aggregate durability testing for airport asphalt in the future.

2 Background

2.1 Airport Asphalt

As stated above, airport asphalt in Australia and many other countries is designed based on the Marshall method (White 1985) with samples compacted by 75 blows of a Marshall hammer. Runway asphalt is generally 14 mm maximum nominal size and is typically constructed in 50–80 mm thick layers. Australian airport asphalt is usually densely graded with a high bitumen content and hydrated lime is often added as an anti-stripping agent (Table 2).

2.2 Aggregate for Asphalt

In practice, the mineral component of airport asphalt mixtures almost always consists of crushed coarse and fine aggregate, natural sand and a hydraulic filler (Zelelew and Papagiannakis 2012). The hydraulic filler is often hydrated lime, although fly ash has also been used from time to time (Liao et al. 2013). The natural sand is usually sourced from rivers or natural sand pits. In practice, 10–15% natural sand is common. Excessive natural sand can lead to deformation prone mixtures while asphalt mixtures with inadequate natural sand are often stiff and unworkable (White 2018).

As detailed above, the aggregate comprises around 93% (by mass) of an airport asphalt mixture. Despite a general focus on the importance of the bituminous binder on airport asphalt performance, the aggregate also plays an important part. However, because most performance related asphalt tests are dominated by the effects of the bituminous binder, aggregate is usually still specified in a prescriptive manner.

Aggregates are routinely characterised by a combination of the consensus properties (angularity, size and shape) as well as their source properties (abrasion resistance, strength, deleterious material content and chemical/mineral composition) (Bessa et al. 2012). The consensus properties are greatly affected by the quarry operation and crushing processes. For example, the shape of aggregate particles can be adjusted by adding tertiary crushing processes. In contrast, the source properties are inherent to the rock and can not be adjusted by processing.

In general, the consensus properties affect how the aggregate fractions are combined to achieve an asphalt mixture with an appropriately graded and interlocking aggregate skeleton, while the source properties are more focussed on the durability of the aggregate and therefore the weathering of the asphalt surface. For example, reactivity to environmentally common chemicals may result in reactions within the aggregate minerals, in turn reducing the asphalt surface life.

2.3 Airport Asphalt Aggregate Specification

Different jurisdictions specify different properties and values for aggregates used in airport asphalt production. However, specifications generally aim to control (CCAA 2009):

- Size and shape.
- Durability.
- Absorptivity.
- Affinity to bitumen.
- Frictional characteristics.
- Contamination.

As stated above, aggregate durability is important to asphalt surface performance, but there is no single and direct test for aggregate durability. Therefore, most jurisdictions use a combination of aggregate properties as indirect indicators of aggregate durability. Table 3 summarises the properties specified for the durability of airport asphalt aggregate in different jurisdictions. Related tests, such as water absorption and plasticity index, are also detailed, although it could be argued that these are not intended to be indicators of aggregate durability. The Australian specification has been criticised by practitioners for containing redundant and excessive durability tests. In fact, in addition to the properties detailed in Table 3, previous versions of the Australian airport asphalt specification also required:

- Unsound and Marginal stone content (AS 1141.30.1). Maximum 1%.
- Methylene blue value, fine aggregate only (AS 1141.66). Maximum 10 mg/g.
- Secondary mineral content (AS 1141.26). Maximum 20%.

The over-specification of airport asphalt aggregate durability in Australia generally reflects the different approaches to road asphalt aggregate specification in each of the Australian States. Each State has its own road asphalt specification and some States prefer different aggregate durability indicators (Table 4). But because the same airport asphalt specification is used across all States, all the durability tests were traditionally included. This contradicts industry advice which recommends any of the following combinations of standard tests to optimise aggregate durability specification (CCAA 2009):

- Wet strength and wet-dry strength variation, or
- Los Angeles abrasion and sodium sulphate soundness, or
- Los Angeles abrasion and unsound/marginal stones.

It is clear that there are various tests and approaches for evaluating the durability of aggregate for asphalt production. It is also clear that some jurisdictions take a significantly more sophisticated approach than recommended by CCAA (2009). What is not clear is whether the various tests provide consistent results. That is, whether a material that is considered unacceptable, or marginal, under one testing regime would also be considered unacceptable, or marginal, under another. Similarly, if a particular aggregate passes one regime but not another, it is not clear whether that material should be accepted, or rejected, or re-tested using an alternate combination of durability indicators.

3 Methods and Results

3.1 Methods

Eight diverse sources of aggregate were sampled and the 10 mm fraction was tested for the Australian airport asphalt specification indicators of course aggregate durability (AAPA 2018):

- Wet strength. According to AS 1141.22.
- Wet-dry strength variation. According to AS 1141.22.

Table 3. Example airport asphalt aggregate durability specifications

Jurisdiction	Specification	Properties	Limit	Test methods
Australia	AAPA (2018)	Los Angeles abrasion	≤30%	AS 1141.23
		Wet strength	≥150 kN	AS 1141.22
		Wet-dry strength variation	≤30%	AS 1141.22
		Sodium sulphate soundness	≤3%	AS 1141.24
		Water absorption	≤2%	AS 1141.6.1
		Plasticity index	Non-plastic (fine only)	AS 1289.3.3.1
United States	FAA (2018)	Los Angeles abrasion	≤40% (course only)	ASTM C131
		Sodium sulphate soundness	≤10/12% (fine/course)	ASTM C88
		Liquid limit	≤25% (fine only)	ASTM D4318
		Plasticity index	≤4% (fine only)	ASTM D4318
United Kingdom	MoD (2009)	Resistance to freeze/thaw	≤18%	BS EN 1367-2
		Los Angeles abrasion	≤30%	BS EN 1097-2
		Water absorption	≤2%	BS EN 1097-6
		Methylene blue value	≤25 km/g (fine only)	BS EN 933-9

- Los Angeles abrasion. According to AS 1141.23.
- Sodium sulphate soundness. According to AS 1141.24.
- Water absorption. According to AS 1141.6.1.

Where available, different sized fractions (7 mm and 14 mm) were tested for the same sources. This allowed the effect of particle size on the test result to be considered. Furthermore, where available, replicate samples of the 10 mm sized fraction were tested, allowing the variability of the testing to be determined.

The sources of aggregate were either used or considered for use in airport asphalt resurfacing projects in Australia (Table 5). The various results were analysed first by looking at the variability of the various tests results for the same aggregate source and nominal fraction size. The difference between the average results for the various nominal sized fractions were also considered, before correlations between the different tests were

Table 4. Australian road asphalt durability specifications

State	Specification	Properties	Limit	Test methods
Victoria	Vicroads (2018)	Degradation factor	Varies with type/use	AS 1141.25.2
		Los Angeles abrasion	Varies with type/use	AS 1141.23
		Unsound/marginal stones	Varies with type/use	AS 1141.30
New South Wales	RMS (2015)	Wet strength	≥150 kN	AS 1141.22
		Wet-dry strength variation	≤35%	AS 1141.22
		Water absorption	≤2.5%	AS 1141.6.1
		Sodium sulphate soundness	≤12%	AS 1141.24
Queensland	TMR (2018)	Degradation factor	≥40%	AS 1141.25.2
		Wet strength	≥150 kN	AS 1141.22
		Wet-dry strength variation	≤35%	AS 1141.22
		Water absorption	≤2.5%	AS 1141.6.1
		Sodium sulphate soundness	≤12%	AS 1141.24
South Australia	DPTI (2018)	Los Angeles abrasion	Project specific	AS 1141.23
		Water absorption	Project specific	AS 1141.6.1
		Sodium sulphate soundness	Project specific	AS 1141.24
		Unsound/marginal stones	Project specific	AS 1141.30
Western Australia	MRWA (2017)	Los Angeles abrasion	≤25%	AS 1141.23
		Wet strength	≥100 kN	AS 1141.22
		Wet-dry strength variation	≤35%	AS 1141.22
		Degradation factor	≥50%	AS 1141.25.2
		Water absorption	≤2%	AS 1141.6.1

Note: Some methods, options and details have been presented in a simplified form.

considered across the various aggregate sources. The results were primarily analysed graphically and using simple statistics, such as mean, standard deviation, coefficient of variation and Student t-tests for differences of means.

Table 5. Aggregate sources

Location	Aggregate type
Alice Springs, Northern Territory	Amphibolite
Archer River, Queensland	Greywacke
Dubbo, New South Wales	Basalt
Mareeba, Queensland	Greywacke
Mildura, Victoria	Basalt
Kununurra, Western Australia	Dolomite
Norfolk Island, New South Wales	Basalt
Rockhampton, Queensland	Greywacke
Proserpine, Queensland	Andesite
Adelaide, South Australia	Dolomite

4 Results

The results for the 10 mm aggregate fractions from each source are summarised in Table 6. The Australian airport asphalt specification limits are also included for reference. The results for the other sized fractions are in Table 7, for the five sources for which data was available. Finally, the replicate 10 mm fraction results are in Table 8, for the three sources for which data was available.

5 Discussion

5.1 Compliance with the Australian Airport Specification

The results for the 10 mm fraction were normalised, such that a value of 1.0 indicated a result at the Australian airport asphalt specification compliance limit and results exceeding 1.0 indicated a source that did not meet the compliance limit. For the aggregate sources considered, wet-dry strength variation, wet strength and water absorption were more likely to result in a source being rejected based on durability indicators (Fig. 1). In fact, all aggregate sources met the Los Angeles abrasion and sodium sulphate soundness requirements. Furthermore, four of the ten aggregate sources met all the Australian airport asphalt durability requirements. This indicates that the sources selected for this research represented the diverse range of materials found in Australia.

Table 6. 10 mm fraction results for all sources

Source	W/D strength (%)	Wet strength (kN)	LA abrasion (%)	SS soundness (%)	Absorption (%)
Specification	≤30	≥150	≤25	≤3	≤2
Alice Springs	45	121	22	0.7	1.6
Archer River	9	147	15	1.3	2.4
Dubbo	23	292	11	0.5	1
Mareeba	21	235	16	1.1	0.8
Mildura	24	133	21	1.9	1.1
Kununurra	13	193	23	0.8	0.3
Norfolk Island	54	139	20	1.1	2.7
Rockhampton	13	376	9	0.2	0.2
Proserpine	15	249	19	0.4	2.5
Adelaide	29	218	20	0.4	2.1

W/D = wet to dry, LA = Los Angeles, SS = Sodium Sulphate.

Table 7. 7 mm and 14 mm fraction results for select sources

Source/Fraction	W/D strength (%)	Wet strength (kN)	LA abrasion (%)	SS soundness (%)	Absorption (%)
Alice Springs					
7	44	111	24	0.2	0.7
14	45	134	19	2.1	1.9
Dubbo					
7	20	354	15	0.9	0.9
14	23	264	13	0.5	1
Mildura					
7	28	134	19	0.5	0.4
14	30	138	13	1.8	1.2
Rockhampton					
7	10	380	17	0.2	0.4
14	11	382	16	0.2	0.2
Proserpine					
7	18	209	13	0.5	0.3
14	15	249	22	0.6	0.9

Table 8. 10 mm replicate fraction results for select sources

Source	W/D strength (%)	Wet strength (kN)	LA abrasion (%)	SS soundness (%)	Absorption (%)
Alice Springs	42	154	19	1.3	0.6
	31	134	18	2.1	0.9
	45	195	18	0.9	0.9
	43	158	23	0.2	0.9
	44	111	24	0.5	0.7
Dubbo	23	264	13	0.5	1
	20	288	15	0.9	0.9
	19	312	9	1.3	1.2
	21	279	14	0.7	1.1
	24	289	16	0.4	0.8
Proserpine	13	262	13	0.8	0.1
	18	244	22	0.9	0.7
	15	266	21	1.0	0.9
	18	239	14	0.7	1.9
	22	209	18	0.5	0.3

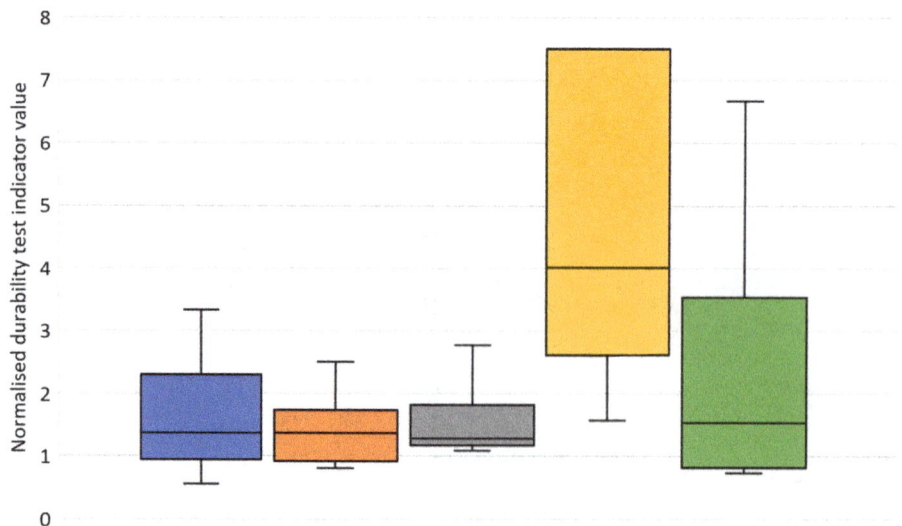

Fig. 1. Normalised 10 mm fractions results

5.2 Variability of Results

The replicate 10 mm fraction results allowed the variability of results for a nominally identical material to be analysed. Table 9 summarises the coefficients of variation for each source and for each durability indicator. Including the results in Table 6, there were a total of six replicate results for each of the three aggregate sources considered. The results indicate that Alice Springs aggregate has greater durability indicator variability than the other sources. However, the Alice Springs Los Angeles abrasion was less variable and the Proserpine water absorption was very high.

Table 9. 10 mm aggregate fraction result coefficients of variation

Source	W/D strength	Wet strength	LA abrasion	SS soundness	Absorption
Alice Springs	12.8%	20.8%	12.9%	71.0%	37.5%
Dubbo	9.1%	5.5%	20.1%	47.0%	14.1%
Proserpine	18.9%	8.3%	20.5%	32.3%	88.3%

Some durability indicator results were significantly different for the different sources, but the variability was generally similar. An example is wet strength, which was significantly lower for Alice Springs than for Dubbo and for Proserpine but with comparable variability (Fig. 2). In contrast, the water absorption results were not significantly different, although the range of the results was much greater for Proserpine than for Alice Springs and for Dubbo (Fig. 3). These contrasts suggest that the expected variability and the average value of the various durability indicators is highly material-specific, supporting the retention of currently specified range of tests.

5.3 Effect of Fraction Size on Results

Despite the durability tests being consensus properties, there are differences in the results for the different fraction sizes presented in Table 7. These differences may simply reflect the natural variation observed within a single fraction size (Table 9) or some other effect associated with relative scale of the aggregate particles compared to the test device, or the imperfections that initiate failures, such as the particle micro-voids that allow water absorption. Although the results varied across the fraction sizes, the differences were generally random. For example, Fig. 4 shows the wet strength being almost uniform across the three fraction sizes for all five aggregate sources. In contrast, the water absorption results were generally consistent for all fraction sizes for Dubbo and Rockhampton, but not for Alice Springs, Mildura and Proserpine (Fig. 5). This indicates that any difference in durability test results associated with the different fraction sizes is more likely to reflect the natural variability in the materials and the testing, rather than some effect of particle size on the test result. However, replicate results for each fraction size are required to allow statistically based conclusions to be drawn.

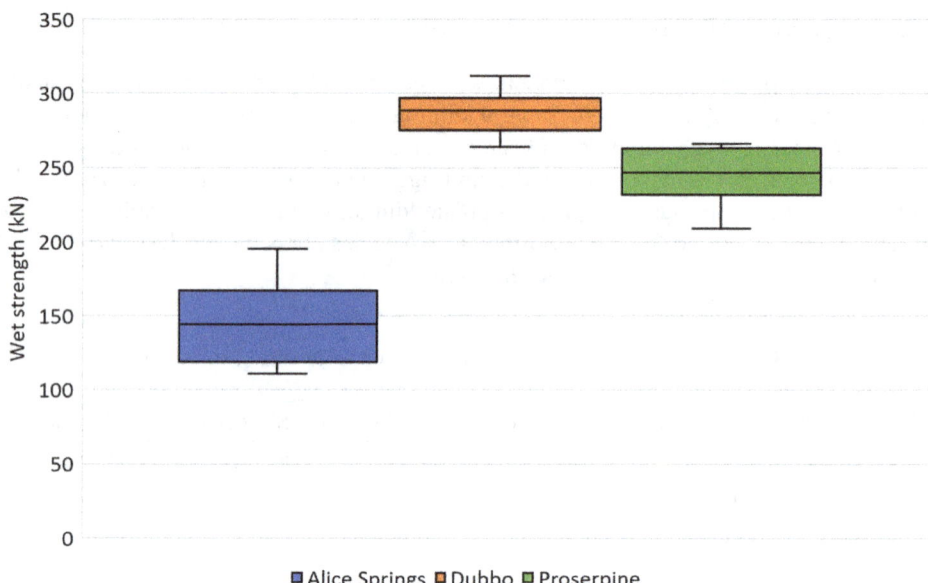

Fig. 2. Distribution of replicate wet strength results

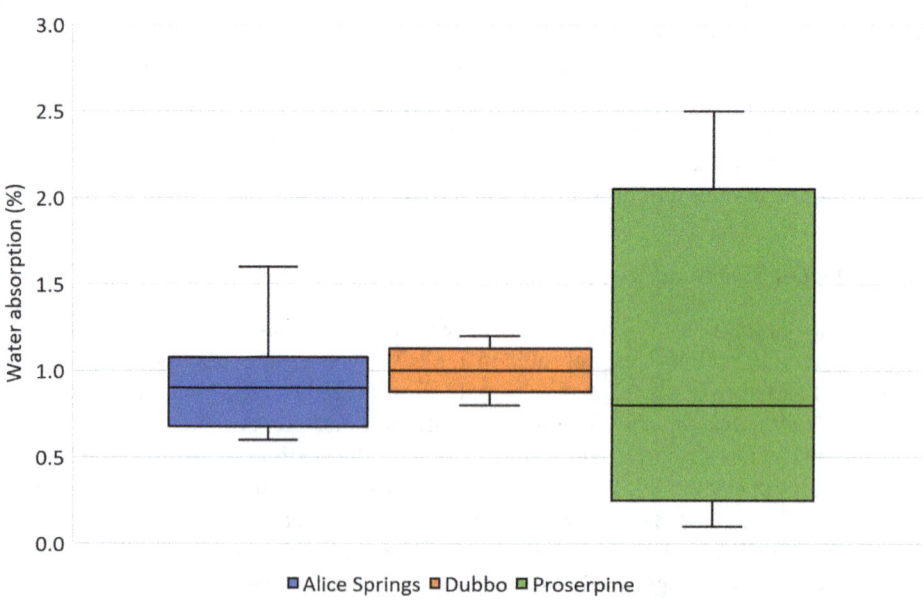

Fig. 3. Distribution of replicate water absorption results

5.4 Relationships Between Durability Indicators

The ability to reduce the range of durability tests for airport asphalt aggregate relies on different indicators providing the same conclusion regarding the suitability of a particular

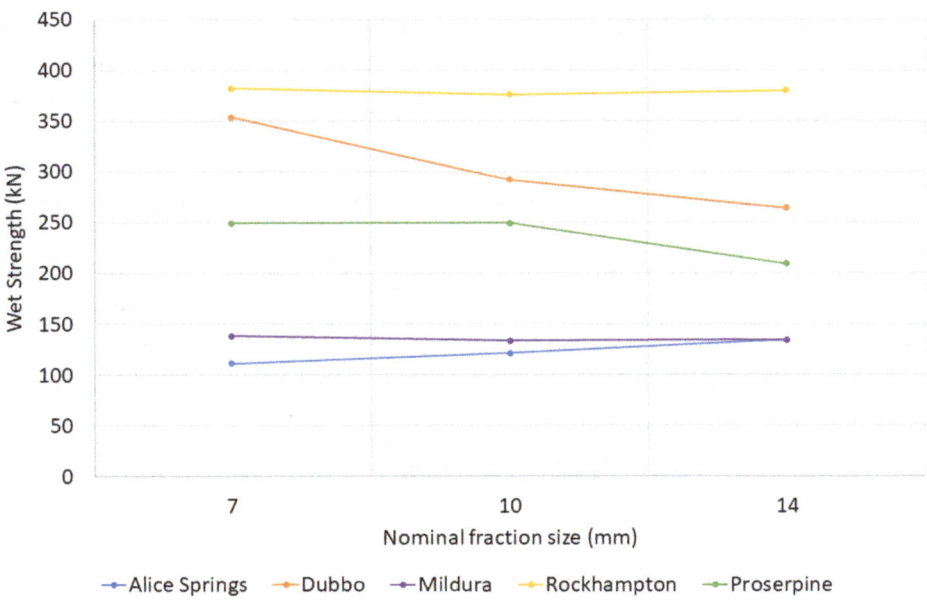

Fig. 4. Wet strength as a function of fraction size

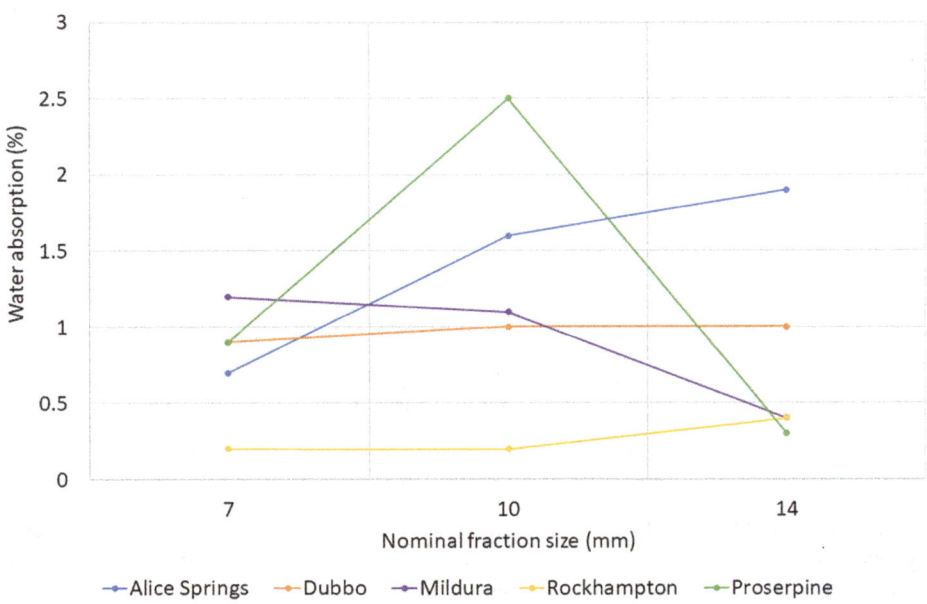

Fig. 5. Water absorption as a function of fraction size

aggregate source. That implies that any redundant tests are highly correlated to each other. Simple linear correlations were developed between the 10 mm fraction results for each durability indicator, across all ten aggregate sources. The resulting coefficients of determination (R^2 values) (Table 10) were less than 0.5, except for the relationship between wet strength and Los Angeles abrasion, which is shown in Fig. 6. In contrast, there was no reportable correlation ($R^2 < 0.01$) between wet-dry strength variation and sodium sulphate soundness (Fig. 7).

Table 10. Coefficients of determination between 10 mm fraction durability test results

Indicator	Wet strength	LA abrasion	SS soundness	Absorption
W/D strength	0.23	0.16	<0.01	0.16
Wet strength	–	0.60	0.49	0.23
LA abrasion	–	–	0.13	0.09
SS Soundness	–	–	–	0.01

All correlations based on first order linear regressions.

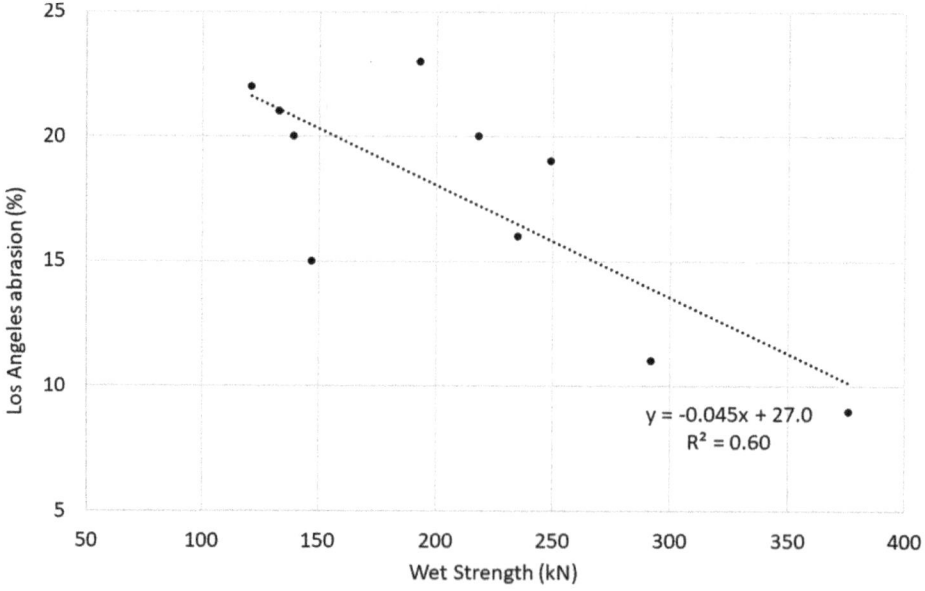

Fig. 6. Relationship between wet strength and LA Abrasion results

The reasonable correlation between wet strength and Los Angeles abrasion is expected to reflect the similar physical mechanisms associated with the two tests. Both involve mechanical damage in the presence of moisture. In contrast, the sodium sulphate test is based on chemical reactivity, while water absorption is a function of the structure of the rock, rather than the minerology of the solid portion of the aggregate particles. It follows that some tests may be redundant and potentially omitted where they test the same physical phenomena, for example mechanical damage in the presence of moisture. However, absorption and chemical reactivity must continue to be tested regardless of the more mechanical test methods.

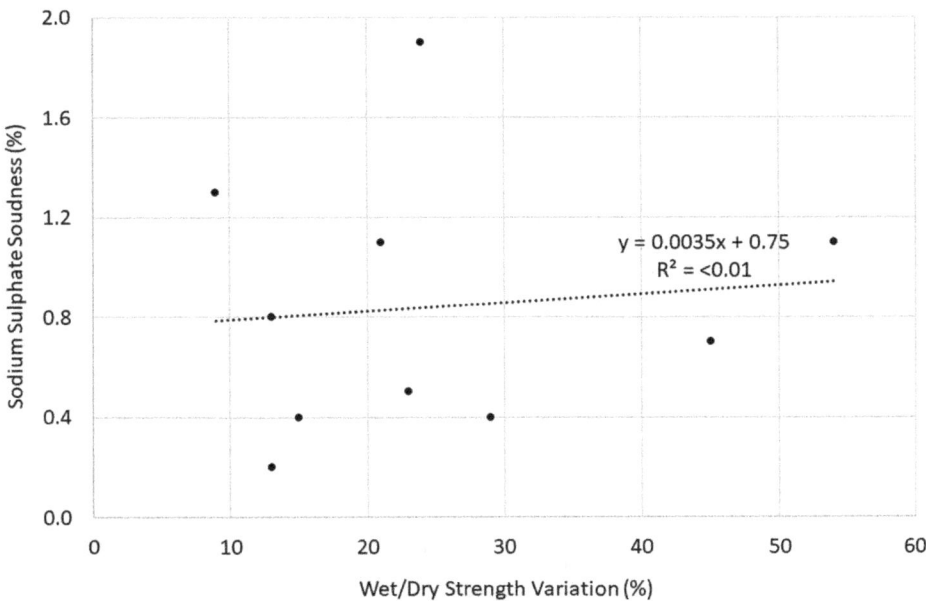

Fig. 7. Relationship between wet-dry strength and sodium sulphate soundness

Water absorption is an interesting test because it is technically not a durability indicator in its own right. However, in practice, high water absorption is anecdotally associated with high wet-dry strength variation. That is because the high absorption allows water to enter the micro-voids within the aggregate particles and this results in reduced mechanical abrasion or crushing resistance. However, very strong aggregates are not affected by the water absorption, as shown in Fig. 8. For example, Proserpine had a high water absorption of 2.5%, but a good wet strength of 249 kN and a wet-dry strength variation of just 15%. Similarly, some aggregates have a low wet strength despite having only modest water absorption, such as Alice Springs, which has 1.6% water absorption but an unacceptable wet strength of just 121 kN and an unacceptable high wet-dry strength variation of 45%.

5.5 Efficacy of Industry Recommended Combinations

As discussed above, industry recommends reducing aggregate durability testing to one of three combinations of durability indicator tests (CCAA 2009). Excluding the combination that includes the percentage of unsound/marginal stones, which is not included in the airport asphalt specification, the recommended combinations are:

- Wet strength and wet-dry strength variation, or
- Los Angeles abrasion and sodium sulphate soundness.

Table 11 summarises the evaluation of each aggregate source against each of the industry recommended combinations. Six sources passed both industry recommended

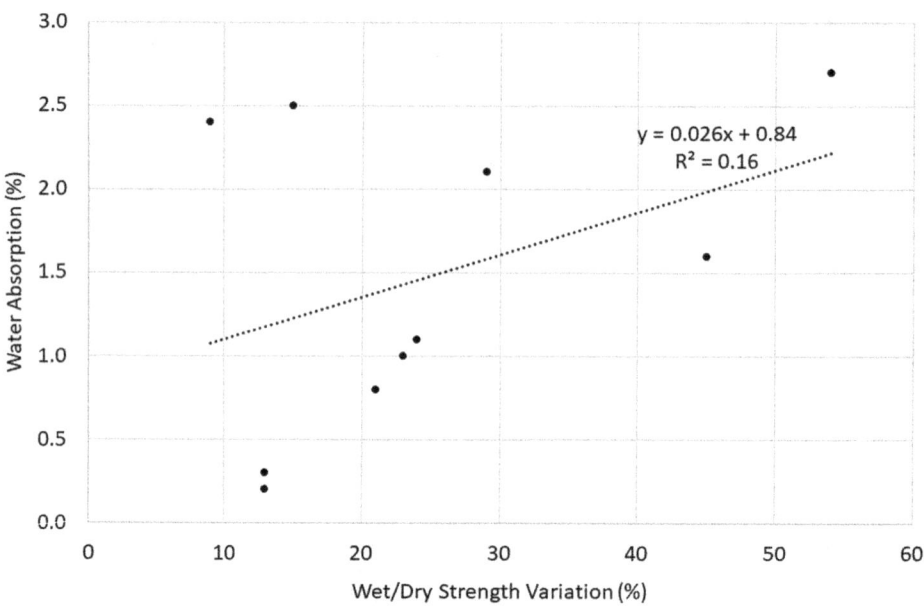

Fig. 8. Relationship between wet-dry strength and absorption

combinations. Of these, two failed the water absorption requirement. Not one source failed on the Los Angeles abrasion and sodium sulphate soundness combination, despite four samples failing the wet strength and wet-dry strength variation combination. This indicates that the Los Angeles abrasion and sodium sulphate soundness combination is not effective and these two industry-recommended combinations are not equivalent. Consequently, the broader combination of durability tests in the current Australian airport asphalt specification can not be replaced by the industry recommended combinations.

Table 11. Acceptance of aggregate based on industry recommended combinations of criteria

Source	Wet strength and wet-dry variation	LA abrasion and SS soundness	Comments
Alice Springs	Fail	Pass	Failed both
Archer River	Fail	Pass	Failed wet strength
Dubbo	Pass	Pass	
Mareeba	Pass	Pass	
Mildura	Fail	Pass	Failed wet strength
Kununurra	Pass	Pass	
Norfolk Island	Fail	Pass	Failed both
Rockhampton	Pass	Pass	
Proserpine	Pass	Pass	
Adelaide	Pass	Pass	

6   Conclusions

It was concluded that wet strength and wet-dry strength variation are the most restrictive of the current aggregate durability requirements in the Australian airport asphalt specification. The different tests are associated with different levels of variability, depending on the aggregate source. Furthermore, the correlation between the various tests results was generally low, except for Los Angeles abrasion and wet strength. The industry recommended combinations of aggregate durability testing were found to be inconsistent and ineffective. Consequently, the current range of aggregate durability tests must be retained. The only exception is the potential to omit Los Angeles abrasion when the wet strength is high. Furthermore, there was no significant difference between the results on the various coarse aggregate fraction sizes and further research is recommended to determine whether acceptable 10 mm fraction results justifies the omission of the testing of the 7 mm and 14 mm fractions from the same source. Further work is required to correlate the various aggregate durability tests to asphalt field performance, although this is expected to be challenging. It would be difficult to isolate the effects of aggregate properties from the overall mixture volumetrics, environmental conditions, traffic loading and bituminous binder properties, when evaluating the field performance of asphalt mixtures. However, a universal, accelerated, laboratory test for the effect of aggregate source properties on the weathering and durability of asphalt mixtures would provide a significant improvement in the future, allowing a more performance-related specification of aggregates for airport asphalt production.

References

AAA: Airfield Pavement Essential, Airport Practice Note 12. Australian Airports Association. Canberra, Australian Capital Territory, Australia, April 2017

Abouelsaad, A., White, G.: Fretting and ravelling of asphalt surfaces for airport pavements: a load or environmental distress? Highways and Airport Pavement Engineering, Asphalt Technology and Infrastructure Conference. Liverpool, England, United Kingdom, 11–12 March 2020

AAPA: Performance-based Airport Asphalt Model Specification. Australian Asphalt Pavement Association. Melbourne, Victoria, Australia, ver. 1.0, January 2018

Bessa, I.S., Branco, V.T.F.C, Soares, J.B.: Evaluation of different digital image processing software for aggregates and hot mix asphalt characterizations. Construct. Build. Mater. **37**, 370–337 (2012)

CCAA: Coarse Asphalt Aggregate: the requirements of AS 2758.5-2009. Cement Concrete and Aggregates Australia. Mascot, New South Wales, Australia (2009)

DPTI: Supply of Pavement Materials, Specification R15. Department of Planning, Transport and Infrastructure. Government of South Australia, July. www.dpti.sa.gov.au/contractor_docu ments/specifications_-_division_R_roadworks. Accessed 28 Dec 2019

Emery, S.: Asphalt on Australian airports. In: Proceedings AAPA Pavements Industry Conference. Surfers Paradise, Queensland, Australia, 18–21 September 2005

FAA: Standard Specifications for Construction of Airfields. AC 150/5370-10H. Federal Aviation Administration, Department of Transportation. Washington, District of Columbia, USA, 21 December 2018

Jamieson, S., White, G.: Improvements to the Australian wheel tracking protocol for asphalt deformation resistance measurement. Aust. Geomech. **54**(2), 113–121 (2019)

Liao, M.-C., Airey, G., Chen, J.-S.: Mechanical properties of filler-asphalt mastics. Int. J. Pavement Res. Technol. **6**(5), 576–581 (2013)

MoD: Marshall Asphalt for Airports. Specification 13. Defence Estates, Ministry of Defence, August 2009

MRWA: Materials for Bituminous Materials, Specification 511. Main Roads Western Australia, Government of Western Australia. 19 August 2019. www.mainroads.wa.gov.au/BuildingRoads/ TenderPrep/Specifications/Pages/500series.aspx. Accessed 28 Dec 2017

Nosler, I., Beckedahl, H.: Adhesion between aggregate and bitumen – performance testing of compacted asphalt specimens by mean of the dynamic indirect tensile test. In: 2nd Eurasphalt and Eurobitume Congress. Barcelona, Spain, 20–22 September 2000

RMS: Aggregates for Asphalt. QA Specification 3152. Roads and Maritime Services, New South Wales Government, 15 July 2015. www.rms.nsw.gov.au/business-industry/partners-suppliers/ document-types/specifications/qa/materials.html. Accessed 28 Dec 2019

TMR: Asphalt Pavements, Specification MRS30. Department of Transport and Main Roads, Queensland Government, March 2018. www.tmr.qld.gov.au/business-industry/Technical-sta ndards-publications/Specifications/5-Pavements-Subgrade-and-Surfacing. Accessed 28 Dec 2019

USACE: Asphalt Surface Airfields PAVER Distress Identification Manual. US Army Corps of Engineers. Vicksburg, Mississippi, United States of America, June 2009

Vicroads: Material Sources for the Production of Crushed Rock and Aggregates. Specification Section 801. Vicroads, Victorian Government, July 2018. webapps.vicroads.vic.gov.au/VRNE/ csdspeci.nsf/. Accessed 28 Dec 2019

White, T.D.: Marshall procedures for design and quality control of asphalt mixture. Asphalt Pavement Technol. **54**, 265–285 (1985)

White, G.: State of the art: asphalt for airport pavement surfacing. Int. J. Pavement Res. Technol. **11**(1), 77–98 (2018)

Zelelew, H.M., Papagiannakis, A.T.: Interpreting asphalt concrete creep behavior through non-Newtonian mastic rheology. Road Mater. Pavement Des. **13**(2), 266–278 (2012)

Author Index

H. Shehata and S. El-Badawy (Eds.): Sustainable Issues in Infrastructure Engineering, SUCI, p. 225, 2021.
https://doi.org/10.1007/978-3-030-62586-3

CPSIA information can be obtained
at www.ICGtesting.com
Printed in the USA
LVHW081759081220
673631LV00001B/7